Handbook of Nanomaterials for Hydrogen Storage

Handbook of Nanomaterials for Hydrogen Storage

edited by
Mieczyslaw Jurczyk

PAN STANFORD PUBLISHING

Published by

Pan Stanford Publishing Pte. Ltd.
Penthouse Level, Suntec Tower 3
8 Temasek Boulevard
Singapore 038988

Email: editorial@panstanford.com
Web: www.panstanford.com

British Library Cataloguing-in-Publication Data
A catalogue record for this book is available from the British Library.

Handbook of Nanomaterials for Hydrogen Storage

Copyright © 2018 Pan Stanford Publishing Pte. Ltd.

All rights reserved. This book, or parts thereof, may not be reproduced in any form or by any means, electronic or mechanical, including photocopying, recording or any information storage and retrieval system now known or to be invented, without written permission from the publisher.

For photocopying of material in this volume, please pay a copying fee through the Copyright Clearance Center, Inc., 222 Rosewood Drive, Danvers, MA 01923, USA. In this case permission to photocopy is not required from the publisher.

ISBN 978-981-4745-66-6 (Hardcover)
ISBN 978-1-315-36444-5 (eBook)

Printed in the USA

Contents

Preface   xi

1. Introduction   1

Mieczyslaw Jurczyk

- 1.1 Motivation   1
- 1.2 Hydrogen   3
- 1.3 Hydrogen Storage   4
- 1.4 Materials Demands   6
- 1.5 Prospects for Nanostructured Metal Hydrides   6

2. Nanomaterials   15

Maciej Tulinski and Mieczyslaw Jurczyk

- 2.1 Introduction   15
- 2.2 Different Approaches to Produce Nanomaterials   16
- 2.3 The Structure of Nanomaterials   18
- 2.4 Methods of Synthesizing Nanomaterials   21
  - 2.4.1 Biological Synthesis   21
  - 2.4.2 Chemical Vapor Deposition   22
  - 2.4.3 Colloidal Dispersion   23
  - 2.4.4 Epitaxial Growth   23
  - 2.4.5 Hydrothermal Synthesis   24
  - 2.4.6 Ion Beam Techniques   25
  - 2.4.7 Lithography   26
  - 2.4.8 Microemulsions   27
  - 2.4.9 Micromachining   27
  - 2.4.10 Physical Vapor Deposition   28
  - 2.4.11 Plasma Synthesis   30
  - 2.4.12 Polymer Route   31

		2.4.13	Pulsed Laser Deposition	31
		2.4.14	Severe Plastic Deformation	32
		2.4.15	Sol-Gel	33
	2.5	Summary		34

3. Solid-State Hydrides   39

Marek Nowak and Mieczyslaw Jurczyk

	3.1	Metal Hydrides		39
	3.2	Intermetallic Hydrides		40
	3.3	Advanced Carbon Hydrides		47
		3.3.1	Fullerenes	47
		3.3.2	Carbon Nanotubes	47
		3.3.3	Graphene	48
	3.4	Complex Hydrides		48
		3.4.1	Alanates	49
		3.4.2	Nitrides and Other Systems	50

4. Preparation Methods of Hydrogen Storage Materials and Nanomaterials   61

Marek Nowak and Mieczyslaw Jurczyk

	4.1	Introduction		61
	4.2	Microcrystalline Hydride Materials		64
	4.3	Nanotechnology for the Storage of Hydrogen		65
		4.3.1	Mechanical Methods	65
	4.4	XPS and EAS Studies		73

5. X-Ray Diffraction   79

Maciej Tulinski

	5.1	Introduction		79
	5.2	Geometry of Crystals		80
		5.2.1	Lattices and Crystal Systems	80
		5.2.2	Designation of Planes	82
		5.2.3	Defects in Crystals	83
	5.3	Properties of X-Rays		85
		5.3.1	Electromagnetic Radiation	85

	5.3.2	The Continuous Spectrum	86
	5.3.3	The Characteristic Spectrum	87
	5.3.4	Production of X-Rays	88
	5.3.5	Detection of X-Rays	89
5.4	Diffraction		92
	5.4.1	Bragg's Law	92
	5.4.2	Laue's Equations	93
	5.4.3	Crystallite Size	95
	5.4.4	X-Ray Diffraction Methods	98

6. Atomic Force Microscopy in Hydrogen Storage Materials Research — 103

Jaroslaw Jakubowicz

6.1	Principles of the AFM Technique	103
6.2	Measurement Procedure	107
6.3	Hydrogen Storage Nanomaterial Imaging	109
6.4	Typical Problems Observed during Nanomaterial Imaging	114
6.5	Conclusion and Future Perspectives	116

7. Characterization of Hydrogen Absorption/Desorption in Metal Hydrides — 119

Mateusz Balcerzak

7.1	What Is a Sievert-Type Apparatus	119
7.2	Preparation of Material to PCT Tests	121
7.3	Types of Tests That Can Be Performed Using Sievert-Type Apparatus	123

8. Electrochemical Characterization of Metal Hydride Electrode Materials — 125

Mateusz Balcerzak and Marek Nowak

8.1	Fundamentals of Electrochemical Research	125
8.2	Preparation of Materials for Electrochemical Measurements	127
8.3	The Results of Electrochemical Measurements	127

9. TiFe-Based Hydrogen Storage Alloys   131

Marek Nowak and Mieczyslaw Jurczyk

  9.1 Phase Diagram and Structure   131
  9.2 Ti–Fe Alloy Synthesized by Mechanical Alloying   133
  9.3 Electronic Structure   138

10. TiNi-Based Hydrogen Storage Alloys and Compounds   149

Mateusz Balcerzak

  10.1 Phase Diagram and Structure   149
  10.2 Electrochemical and Gaseous Hydrogen Sorption Measurements   152
      10.2.1 Cycle-Life Curves   153
      10.2.2 Electrodes and Electrochemical Measurement Conditions   157
      10.2.3 The Influence of Temperature on Hydrogen Sorption/Desorption Properties   158
  10.3 Arc Melted Alloys   159
      10.3.1 Chemical Modification of Arc Melted Alloys   159
      10.3.2 Composites Containing Arc Melted Alloys   161
  10.4 Mechanically Alloyed Alloys   161
  10.5 Other Methods of Alloy Production   168
  10.6 Gaseous Hydrogen Sorption and Desorption of Ti–Ni Alloys   169
  10.7 Electrochemical and Gaseous Hydrogen Sorption and Desorption of Ti_2Ni Chemically Modified by Pd and Multi-Walled Carbon Nanotubes   170

11. ZrV_2-Based Hydrogen Storage Alloys   179

Marek Nowak and Mieczyslaw Jurczyk

  11.1 Zr–V Phase Diagram and Structure   179
  11.2 ZrV_2 Type Alloys Synthesized by Mechanical Alloying   181
  11.3 Electrochemical and Thermodynamic Properties   184
  11.4 Electronic Structure   190
  11.5 Zr-Based Alloys with MWCNT Addition   192

12. LaNi$_5$-Based Hydrogen Storage Alloys — 199
Marek Nowak and Mieczyslaw Jurczyk

 12.1 Phase Diagram and Structure 199
 12.2 LaNi$_5$-Type Compounds 201
 12.3 LaNi$_5$ Phase Synthesized by Mechanical Alloying 203
 12.4 XPS and AES Studies 206
 12.5 Thermodynamical Properties 211
 12.6 Electrochemical Properties 212
 12.7 Electronic Structure 214
 12.8 Composite LaNi$_5$-Type Materials 218

13. Mg-3d-Based Hydrogen Storage Alloys — 227
Marek Nowak and Mieczyslaw Jurczyk

 13.1 Introduction 227
 13.2 Mg–Ni and Mg–Cu Phase Diagrams 228
 13.3 Mg$_2$Ni-Type Alloys 229
 13.4 Mg$_2$Cu-Type Alloys 234
 13.5 Effect of Ball-Milling with Graphite and Palladium 238
 13.6 Amorphous 2Mg + 3d Alloys Doped by Nickel Atoms (3d = Fe, Co, Ni, Cu) 240
 13.7 XPS Valence Band and Segregation Effect in Nanocrystalline Mg$_2$Ni Materials 246
 13.8 Mg-Based Nanocomposite Hydrides for Room Temperature Storage 252

14. (La, Mg)$_2$Ni$_7$-Based Hydrogen Storage Alloys — 261
Marek Nowak and Mieczyslaw Jurczyk

 14.1 Phase Diagram and Structure 261
 14.2 Electrochemical Properties 263
 14.3 (La,Mg)$_2$Ni$_7$-Type Alloys Synthesized by Mechanical Alloying 267
 14.4 RE–Mg–Ni-Based Alloy Electrodes 272

15. Ni-MH$_x$ Batteries — 279
Mieczyslaw Jurczyk and Marek Nowak

 15.1 Introduction 279

15.2 The Fundamental Concept of Hydride Electrode
and Ni-MH$_x$ Battery   280
15.3 Electrode Materials for Ni-MH$_x$ Batteries   282
15.4 Sealed Ni-MH$_x$ Batteries   289
15.5 A Composite Hydrogen Storage Alloy in Application
in Sealed Ni-MH$_x$ Batteries   294
15.6 Major Markets for Ni-MH$_x$ Batteries   295

Index   303

Preface

The most abundant element in the universe is hydrogen. On the earth, it can be found mostly chemically bounded with oxygen, in the form of H_2O. This element is a highly volatile, clean, synthetic (e.g., produced by the dissociation of H_2O) and nontoxic fuel. Combustion of hydrogen does not produce CO_2.

Because of the need to seek alternatives to reducing world fossil fuels, hydrogen is considered a new, renewable fuel. It is a key issue of the hydrogen energy economy to find a safe and efficient system for H_2 storage (under reasonable temperature and pressure). A proper hydrogen storage unit for mobility and transfer application should also have good absorption and desorption kinetics, high storage capacity, practical thermal properties, stable cycling behavior, and low cost.

Alloys and intermetallic compounds that react with hydrogen to form metal hydrides are the most promising candidates for hydrogen storage applications. Also, other systems such as alanates, borohydrides, amide borohydrides, amide–imide, amineborane, and alane for hydrogen storage have been studied. With the rapid development of nickel/metal-hydride (Ni-MH$_x$) batteries, the demand for novel electrode materials with much higher hydrogen storage capacity and better capacity retaining rate is urgent.

For a long time, metal hydrides (MH$_x$) have been in focus as one of several alternatives to store hydrogen in a hydrogen-based economy. A large number of hydrogen storage materials have been studied until now. As a result, many hydride alloys (TiNi, TiFe, LaNi$_5$) were studied and developed. Also, AB$_2$-type Ti/Zr-V-Ni-M systems and rare earth–based AB$_5$-type La–Ni–Co–Mn–Al alloys were investigated. These systems obtained long-life hydrogen storage electrode alloys, but the Ti/Zr system has a higher capacity. The non-stoichiometric AB$_2$ and AB$_5$ systems with improved performance such as higher capacity and long cycle life were developed. Some example compounds include LaNi$_5$H$_6$ and Mg$_2$NiH$_4$, which have hydrogen densities currently reported at 1–3.5%.

The search continues for new hydrogen storage materials, both conventional metal hydrides and new materials such novel compounds with a high hydrogen contents and sufficiently fast reaction kinetics of absorption and desorption at moderate temperatures. Hydrogen storage materials research has entered a new and exciting period with the advance of the nanocrystalline alloys, which show substantially enhanced absorption/desorption kinetics, even at room temperature.

Nanotechnology involves the precise manipulation and control of atoms, the building elements of all matter, to create new materials. It is widely accepted that this technology is developing into a major driver for commercial success in the 21st century. Over the past decade, the use of nanostructured metallic materials has already changed the approach to materials design in many applications, by seeking structural control at the atomic level and tailoring of the physico-chemical, mechanical engineering and biological properties. Today, it is possible to prepare metal nanocrystals with nearly monodispersive size distribution. Nanomaterials demonstrate novel properties compared with conventional (microcrystalline) materials due to their nanoscale features.

Hydrogen storage nanomaterials can be produced by many methods, one of which is mechanical alloying (MA). This method consists of repeated fracturing, mixing, and cold welding of a fine blend of elemental particles, resulting in size reduction and chemical reactions. MA has been recently used to produce nanocrystalline, non-equilibrium metal hydride materials. It has already been observed that the kinetic barriers in nanocrystalline hydrides should be low compared with coarse-grained materials. Moreover, the MA process increases diffusion channels and obviously shortens diffusion paths for H-atoms. There are numerous possibilities for the design, synthesis, and control of the properties of nanostructured, multicomponent hydrogen storage materials and nanocomposites fabricated by the application of the MA process.

By making the metal storage material in the form of nanoparticles, two advantages can be gained, one related to the kinetics and the other to the thermodynamics. The kinetics aspect is obvious; with nanoparticles, the uptake and release kinetics become much faster than with the more commonly used microcrystalline particles. The second aspect concerns the possibility to control the thermodynamic properties by tuning particle size. Since the thermodynamic properties of sufficiently small particles change due to, for instance, surface energy and elasticity/plasticity effects, the dissociation pressure may be varied for a given material by using particle size as a tuning parameter.

Recently, the MA process for the fabrication of alloys and their nanocomposites with a unique microstructure has been developed. The process permits the control of microstructural properties and the nature of the base alloy or composite. The availability of large amounts of specifically tailored nanostructured metal-based powders is crucial for the successful development of new energy materials for the hydrogen storage.

The present research aims to fabricate nanostructured hydrogen storage materials and their nanocomposites. Several factors (crystal structure, microstructure, crystallite size, purity of the produced nanomaterials, electronic structure) have shown their influence on the final properties, such as hydride stability, hydrogen storage capacity, thermodynamics and kinetics of hydrogenation–dehydrogenation processes, reversibility and hydrogen storage capacity, solid phase/gas and solid phase/electrolyte solution systems and finally on electrochemical properties.

This book is our contribution to this innovative area of nanomaterials and nanocomposites for hydrogen storage applications. Wherever possible, we illustrate the subject being discussed with the help of our own results. The content of this book is arranged into several chapters. The first chapter emphasizes the motivation for the transformation to the nanomaterials and synthesis of nanomaterials, aiming at describing the principles and approaches of the synthesis techniques. In a number of following sections, we provide a comprehensive history of the development of materials, including the existing fabrication methods, with a special emphasis on MA or high-energy ball-milling in high-energy mills. There are many different ways to produce nanomaterials (Chapter 2). The traditional and simpler way of producing fine particles has been "top-down," referring to the reduction through attrition and various methods of comminution in the traditional sense. Nowadays, methods of production using "bottom-up" techniques are being increasingly utilized. There is a renewed interest in solid-state hydrides, both conventional metal hydrides (Chapter 3) and new materials such as carbon nanostructures and novel compounds. In Chapter 4, we review preparation methods of hydrogen storage materials and nanomaterials. The following chapter present a thorough review of the X-ray diffraction (Chapter 5) and atomic force microscopy (Chapter 6) in hydrogen storage materials research as well as the characterization of hydrogen absorption–desorption in metal hydrides (Chapter 7) and electrochemical characterization of metal hydride electrode materials (Chapter 8). The other parts of the book describe TiFe-based hydrogen storage alloys (Chapter 9), TiNi-based hydrogen storage alloys (Chapter 10), ZrV_2-based hydrogen storage alloys (Chapter 11), $LaNi_5$-based hydrogen storage alloys (Chapter 12), Mg-3d-based hydrogen storage alloys (Chapter 13), and $(La, Mg)_2Ni_7$-based hydrogen storage alloys (Chapter 14). The final part (Chapter 15) focuses on the application of bulk nanostructured materials in Ni-MH$_x$ rechargeable alkaline cells.

The potential application of these research fits well to the EU Framework Programme for Research and Innovation Horizon 2020, where one of the societal challenges is secure, clean, and efficient energy. The replacement of conventional technologies with hydride technologies may also contribute to the reduction of greenhouse gas emissions.

Our goal is to provide comprehensive and complete knowledge about materials for energy applications to the graduate students and researchers with a background in chemistry, chemical engineering, and materials science.

I express my appreciation to all of the authors for their contributions.

Mieczyslaw Jurczyk
Poznan University of Technology, Poland

Chapter 1

Introduction

Mieczyslaw Jurczyk

*Institute of Materials Science and Engineering,
Poznan University of Technology, Poznan, Poland*

mieczyslaw.jurczyk@put.poznan.pl

1.1 Motivation

The worldwide population and economic growth is increasing the energy demand at a dramatic pace. Today, combustion of fossil fuels provides 86% of the world's energy. Drawbacks to fossil fuel utilization include limited supply, pollution, and carbon dioxide emissions. Carbon dioxide emissions, thought to be responsible for global warming, are now the subject of international treaties. Together, these drawbacks argue for the replacement of fossil fuels with a less-polluting energy carrier, such as hydrogen. Hydrogen has great potential as an environmentally clean energy fuel and is the simplest and most abundant of the chemical elements, constituting roughly 75% of the universe's elemental mass. Hydrogen fuel is approaching the edge of surging into industry, transportation, and home us as an accessible source of energy [1–12]. Increasing application of hydrogen energy is the only way forward to meet the objectives of US Department of Energy (DOE), i.e., reducing green house gases, increasing energy

---

Handbook of Nanomaterials for Hydrogen Storage
Edited by Mieczyslaw Jurczyk
Copyright © 2018 Pan Stanford Publishing Pte. Ltd.
ISBN 978-981-4745-66-6 (Hardcover), 978-1-315-36444-5 (eBook)
www.panstanford.com

security and strengthening the developing countries economy [13]. Hydrogen has been projected as a long-term solution for a secure energy future.

The production of hydrogen is an appropriate environmental solution. Hydrogen, as a carbon-less fuel, forms no carbon dioxide or particulates during combustion. While carbon dioxide emission from vehicles is currently not regulated, the increased air and consequent temperature needed to efficiently burn particulates formed during the combustion of carbon-containing fuels causes NO_x formation. This difficult balance between particulate and NO_x formation is avoidable when hydrogen is the fuel of choice. Another advantage of hydrogen as an energy carrier is its flexibility. Localized production of hydrogen is feasible nearly everywhere from several sources. Currently, the primary route for hydrogen production is the conversion of natural gas and other light hydrocarbons [14]. Coal and petroleum coke may also serve as raw materials for hydrogen production in the future. On the other hand, the clean way to produce hydrogen from water is to use sunlight in combination with photovoltaic cells and water electrolysis.

Nanotechnology could help speed up the journey to the hydrogen society [15–37]. Nanotechnology's major contribution to the clean production of hydrogen lies in its application to solar cells and the catalysts used in water electrolysis. It has been shown that nanoscale electrode materials will increase the efficiency of the cell. Storing the hydrogen onboard that is needed to run your car's fuel cells poses another challenge. There are three ways of doing this: as a high-pressure compressed gas, a cryogenic liquid or as a solid. Many metals are capable of absorbing hydrogen as well. Nanotechnology plays an important role also here [5, 12]. Nanomaterials have diverse tunable physical properties as a function of their size and shape due to strong quantum confinement effects and large surface to volume ratios. These properties are useful for designing hydrogen storage materials. For instance, researchers are now investigating nanostructured polymeric materials as hydrogen storage adsorbents [38]. The new polymer adsorbent material has shown great promise.

Due to their large surface areas with relatively small mass, single-walled carbon nanotubes (SWCNTs) have been considered very promising potential materials for high capacity hydrogen storage, as well [7, 39–45]. Theoretically, they can store hydrogen up to 7.7 wt%, as every carbon atom in SWCNTs chemisorbs one hydrogen atom. Generally the BET surface area of CNTs ranges between 50–1315 m^2/g and it changes by the producing method, adsorbate gas used for measurements and the specific structure and composition of samples tested. There are many methods in the literature about increasing the specific surface area of CNTs and most of them are also used to purify the material, such as wet techniques depend on the activation with commercially available acids or bases and oxidation of CNTs. It is known that gases could be adsorbed inside the tubes, in the space between the tubes, or only on the outer surface of bundle. The opening of the tubes by the removal of the caps is another way that increases adsorption in nanotubes. It was also reported that heat treatment removed the sorbed hydrocarbons that were present in the sample.

Mobile applications in combination with hydrogen fuel cell systems require sustainable storage materials which contain large amounts of hydrogen. It has been shown that light metal hydrides show a high potential for reversible hydrogen storage applications [1, 7, 46–49]. The hydrogen content reaches values of up to 9.3 wt% for magnesium alanate, $Mg(AlH_4)_2$. While the hydrogen content of these compounds is high and desorption kinetics seems to be promising, the absorption kinetics still has to be improved. Additionally, it has already been observed that the kinetic barriers in nanocrystalline hydrides should be low compared to coarse-grained materials [50]. The search continues for new materials with a high hydrogen contents and sufficiently fast reaction kinetics of absorption and desorption at moderate temperatures.

1.2 Hydrogen

Hydrogen is the lightest of all elements and the most abundant one in the universe [51]. Hydrogen gas is transparent, odorless, and nontoxic. On earth, however, it only ranks the 15th element, and its concentration in the atmosphere is very low (0.5 ppm). Yet it forms the largest number of chemical compounds, water being the most abundant of them.

In its gaseous state, hydrogen forms the diatomic molecule H$_2$ (dihydrogen). Some basic properties of hydrogen are listed in Table 1.1. It can be seen that it is a solid below −260.85°C, and its liquid form exists only in a small range of temperatures up to −253.25°C. Above its critical point (−240°C; = 1.276 MPa) hydrogen is a supercritical fluid.

Table 1.1 Hydrogen properties

Atomic weight	1.0079	g mol^{-1}
Van der Waals radius	120	pm
Covalent radius	31	pm
ΔH (H$_2$ → 2 H)	436	kJ mol^{-1}
Density at 0°C and 0.1 MPa	0.0899	g l^{-1}
Liquid density (−253.25°C)	70.8	g l^{-1}
Boiling point	20.25	K
Critical temperature	32.97	K
Critical pressure	1.276	MPa
Self-ignition temperature	474	°C
Flammability limit in air	4.1–74.2	Vol.%
Explosive limits in air	18.3–59	Vol.%
Diffusion coefficient in air	0.634	cm^2 s^{-1}

Hydrogen can be stored, either in tank (e.g., in the case of high-pressure hydrogen storage or liquefied hydrogen storage) or in a tank containing a solid (solid-state H$_2$ storage). In the latter case, the storage density is often referred exclusively to the solid used, in order to facilitate the performance comparison of different hydrogen storage materials/nanomaterials.

1.3 Hydrogen Storage

Hydrogen storage is a topical goal in the development of a hydrogen economy. The methodologies span many approaches, including high pressures and cryogenics, but usually focus on chemical compounds that reversibly release H$_2$ upon heating [1, 51].

This third possibility is to store hydrogen through dissociative chemisorption, followed by the formation of a new compound (metallic hydrides, MH$_x$). During chemisorption in metal hydrides, the hydrogen molecule is first dissociated on the surface, and then its atoms diffuse into the metal host. Inside the bulk, chemical bonds are formed, resulting in the hydride phase. With respect to the bonding mechanism between the hydrogen and the host material, metal hydrides can be categorized, at a first glance, as saline hydrides (ionic bonding), covalent hydrides (covalent bonding), and interstitial hydrides (metallic bonding).

Most research into hydrogen storage is focused on storing hydrogen in a lightweight, compact manner for different applications. For the hydrogen economy to take hold, there must be significant technological breakthroughs that allow hydrogen energy to be harnessed on a small scale by vehicles and appliances. For hydrogen storage, some targets have been established (Table 1.2) [52]. They will allow for its use hydrogen in a fuel cell vehicle.

Table 1.2 Technical targets: onboard hydrogen storage systems

Storage parameter	Units	2010	2015	Ultimate target
Gravimetric system density	kWh kg^{-1}	1.5	1.8	2.5
	MJ kg^{-1}	5.4	6.48	9
	wt%	4.5	5.5	7.5
Volumetric system density	kWh l^{-1}	0.9	1.3	2.3
	MJ m^{-3}	3.24	4.68	8.28
	l^{-1}	28	40	70
Fuel cost at pump	$	3–7	2–6	2–3
System filling time for 5 kg H$_2$	min	4.2	3.3	2.5
	min^{-1}	1.2	1.5	2
Maximum loss of useable H$_2$	g (h)$^{-1}$	0.1	0.05	0.05

Note: 1 gasoline gallon equivalent (GGE) 1 kg hydrogen.

Onboard hydrogen storage for transportation applications continues to be one of the most technically challenging barriers to the widespread commercialization of hydrogen-fueled light-duty vehicles. The U.S. Department of Energy's (DOE's) Office of Energy Efficiency and Renewable Energy (EERE) and Fuel Cell Technologies (FCT) Program's hydrogen storage activity focuses primarily on the applied research and development (R&D) of low-pressure, materials-based technologies to allow for a driving range of greater than 300 miles (500 km) while meeting packaging, cost, safety, and performance requirements to be competitive with comparable vehicles in the market place.

Current technologies may need to improve by a factor of two or more in order to meet some of these needs. The sections below detail the different approaches for storing hydrogen. High-pressure storage still remains the most practical option for storage. A key drawback of hydrogen gas is that it has an extremely low density.

1.4 Materials Demands

Designing an optimized materials and nanomaterials hydrogen storage is a challenging task [53]. Generally two basic requirements should be taken into account: (i) a large number of adsorption sites per weight and volume of hydrogen storage materials and (ii) the heat of adsorption of the hydrogen storage materials should be increased. For example, at room temperature, the optimum adsorption enthalpy of hydrogen storage material should be between 10 and 50 kJ and theoretical calculations reveal a value of 15.1 kJ.

The combined demands of reversibility of hydrogen loading, high capacity, low volume, low weight, price, safety, and ease of operation mean that currently no material yet meets the constraint for reversible hydrogen storage under near-ambient conditions.

1.5 Prospects for Nanostructured Metal Hydrides

A few currently available technologies permit to store directly hydrogen by modifying its physical state in gaseous or liquid form. However, these methods are not able to meet all of the criteria proposed below [5–56]:

- high hydrogen content per unit mass and unit volume
- limited energy loss during operation

- fast kinetics during charging
- high stability with cycling
- cost of recycling and charging infrastructures
- safety concerns in regular service or during accidents

In order to optimize the choice of the intermetallic compounds for a battery application, a better understanding of the role of each alloy constituent on the electronic properties of the material is crucial. The semi-empirical models showed that the energy of the metal—hydrogen interaction depends both on the geometric as well as electronic factors [57–59]. The nanocrystalline metal hydrides offer a breakthrough in prospects for practical applications. Their excellent properties (significantly exceeding traditional hydrides) are a result of the combined engineering of many factors: alloy composition, surface properties, microstructure, grain size, and others. In the development of nanocrystalline hydrides, the goal is not only to improve operational properties of the existing hydrides, but also (more importantly) to create a new generation of materials, with the properties being designed and controlled to fulfill the particular demands of different applications.

Nanostructured metal hydrides are a new class of materials in which outstanding hydrogen sorption may be obtained by proper engineering of the microstructure and surface [5, 12, 13, 21–23, 25–29, 31, 32, 35, 37]. The case of these materials is extremely important, because it can be applied to hydrogen storage systems and Ni-MH$_x$ batteries. The techniques of mechanical alloying and high-energy ball milling have been successfully used to improve the hydrogen sorption properties of various metal hydrides. Nanocrystalline powder alloys readily absorb hydrogen with no need for prior activation in contrast to conventional hydrides. These materials show substantially enhanced absorption and desorption kinetics even at relatively low temperatures. For the vehicle application, depending on the temperature of hydrogen absorption/desorption below or above 150°C, the alloy hydrides can be distinguished in high and low temperature materials. The principal disadvantages of alloy hydrides, apart from the cost, are the low hydrogen content at low temperature (e.g., La-based alloys) and the difficulty of reducing desorption temperature and pressure of alloy hydrides having high hydrogen storage capacity and fast rated (e.g., Mg-based materials). To solve the above-mentioned problems, the use of composite materials, starting from Mg-alloys, and of novel catalyzed metal hydrides (ZrV_2, $LaNi_5$, TiFe) have been studied [60].

Nanocomposite materials possess excellent hydrogen storage characteristics. Use of such composite materials with the dominant component as the main matrix and the other minority components as secondary phase, may lead to better hydrogen storage property due to the interface effects arising from significant density of interface boundaries. It has been reported that ball milling of Mg with transition metals, as well as intermetallics or carbon nanostructures improved hydrogen sorption kinetics [61–65].

Recently, La–Mg–Ni compounds with A_2B_7 type structure are already utilized in novel Ni-MH$_x$ batteries [66–70]. This system (A_2B_7-type) is considered to be the most promising candidates as the negative electrode materials of Ni-MH$_x$ rechargeable battery. Within 30 cycles, the $La_5Mg_2Ni_{23}$ alloy electrode, showed perfect cyclic stability. Discharge capacity for this alloy was 410 mAh/g [66, 67].

In order to optimize the choice of the nanocomposites for a selected application, a better understanding of the role of each alloy constituent on the electronic properties of the material is crucial. Our studies at Poznan University of Technology (Poland) may supply useful indirect information about the influence of the valence band structure, surface chemical composition, crystal structure, grain sizes, and preparation conditions on the hydrogenation properties of the various hydrogen storage nanomaterials as well as nanocomposites.

The nanocrystalline metal hydrides offer a breakthrough in prospects for practical applications. In the development of nanocrystalline hydrides, the goal is not only to improve operational properties of the existing hydrides, but also (more importantly) to create a new generation of materials, with the properties being designed and controlled to fulfill the particular demands of different applications.

References

1. Jain, I. P. (2009). Hydrogen the fuel for 21st century. *Int. J. Hydrogen Energy* **34**: 7368–7378.

2. Hodes, G., ed. (2001). *Electrochemistry of Nanomaterials*. WILEY-VC.
3. Grégoire Padró, C. E., and Lau, F., eds. (2002). *Advances in Hydrogen Energy*, Kluwer Academic Publ.
4. Jones, R. H., and Thomas, G. J., eds. (2008). *Materials for the Hydrogen Economy*, CRC Press.
5. Varin, R. A., Czujko, T., and Wronski, Z. S. (2009). *Nanomaterials for Solid State Hydrogen Storage*, Springer Science+Business Media, LLC.
6. Sakintuna, B., Lamari-Darkrim, F., and Hirscher, M. (2007). Metal hydride materials for solid hydrogen storage: A review. *Int. J. Hydrogen Energy* **32**: 1121–1140.
7. Jain, I. P., Jain, P., and Jain, A. (2010), Novel hydrogen storage materials: A review of lightweight complex hydrides. *J. Alloys Comp.* **503**: 303–339.
8. Zhao, X. G., and Ma, L. Q. (2009). Recent progress in hydrogen storage alloys for nickel/metal hydride secondary batteries—review. *Int. J. Hydrogen Energy* **34**: 4788–4796.
9. Bououdina, M., Grant, D., and Walker, G. (2006). Review on hydrogen absorbing materials-structure, microstructure, and thermodynamic properties. *Int. J. Hydrogen Energy* **31**: 177–182.
10. Young, K. H., and Nei, J. (2013). The current status of hydrogen storage alloy development for electrochemical applications. *Materials* **6**: 4574–4608.
11. Principi, G., Agresti, F., Maddalena, A., and Lo Russo, S. (2009). The problem of solid state hydrogen storage. *Energy* **34**: 2087–2091.
12. Jurczyk, M., Smardz, L., and Szajek, A. (2004). Nanocrystalline materials for NiMH batteries. *Mater. Sci. Eng. B* **108**: 67–75.
13. http://energy.gov/sites/prod/files/2013/12/f5/Draft%20DOE%20Strategic%20Plan%2012-4-13%20for%20Public%20Comment%20FINAL.pdf—"2014–2018 DOE Strategic Plan."
14. Winter, C. J., and Nitsch, J. (1988). *Hydrogen as an Energy Carrier: Technologies*, Systems, Economy, Springer.
15. Dowling, A. P. (2004). Development of nanotechnologies, *Mater. Today*, **7**: 30–35.
16. Wang, Z. L., Liu, Y., and Zhang, Z., eds. (2003). *Handbook of Nanophase and Nanostructured Materials*, Kluwer Academic/Plenum Publishers, New York, chapter 1.
17. Yeadon, M., Yang, J. C., Ghaly, M., Olynick, D. L., Averbach, R. S., and Gibson, J. M. (1997). *Nanophase and Nanocomposite Materials,*

Komarneni, S., Parker, J. C., and Wollenberger, H. J., eds., MRS, Pittsburgh, PA, pp. 179–184.
18. Benjamin, J. S. (1976). Mechanical alloying. *Sci. Am.*, **234**: 40–57.
19. Suryanarayna, C. (2001). Mechanical alloying. *Progr. Mater. Sci.*, **46**, pp. 1–184.
20. Suryanarayana, C., and Koch, C. C. (1999). Nanostructured materials, in *Non-Equilibrium Processing of Materials*, Suryanarayana, C., ed. (Elsevier Science Pub., Oxford, UK), pp. 313–346.
21. Lü, L., and Lai, M. O. (1998). *Mechanical Alloying* (Boston, London-Kluwer Academic).
22. Jurczyk, M., and Nowak, M. (2008). Nanomaterials for hydrogen storage synthesized by mechanical alloying, in: *Nanostructured Materials in Electrochemistry*, Ali, E., ed., Chapter 9, (Wiley).
23. Balcerzak, M., Jakubowicz, J., Kachlicki, T., and Jurczyk, M. (2015). Hydrogenation properties of nanostructured Ti_2Ni-based alloys and nanocomposites. *J. Power Sources* **280**: 435–445.
24. Varin, R. A., Chiu, Ch., and Wronski, Z. S. (2008). Mechano-chemical activation synthesis (MCAS) of disordered $Mg(BH_4)_2$ using $NaBH_4$. *J. Alloys Comp.* **462**: 201–208.
25. Benjamin, J. S. (1997). Mechanical alloying process (United States Patent 5688303).
26. Li, X. D., Elkedim, O., Nowak, M., and Jurczyk, M. (2014). Characterization and first principle study of ball milled Ti-Ni with Mg doping as hydrogen storage alloy. *Int. J. Hydrogen Energy* **39**: 9735–9743.
27. Huang, L. W. S., Elkedim, O., Nowak, M., Jurczyk, M., Chassognon, R., and Meng, D. W. (2012). Synergistic effects of multiwalled carbon nanotube and Al on the electrochemical hydrogen storage properties of Mg_2Ni-type alloy prepared by mechanical alloying. *Int. J. Hydrogen Energy* **37**: 1538–1545.
28. Smardz, L., Jurczyk, M., Smardz, K., Nowak, M., Makowiecka, M., and Okonska, I. (2008). Electronic structure of nanocrystalline and polycrystalline hydrogen storage materials. *Renew. Energy* **33**: 201–210.
29. Smardz, K., Smardz, L., Okonska, I., Nowak, M., and Jurczyk, M. (2008). XPS valence band and segregation effect in nanocrystalline Mg_2Ni-type materials. *Int. J. Hydrogen Energy*, **33**: 387–392.
30. Smardz. L., Smardz, K., Nowak, M., and Jurczyk, M. (2001). Structure and electronic properties of $La(Ni,Al)_5$ alloys. *Cryst. Res. Tech.*, **36**: 1385–1392.
31. Courtney, T. H., and Maurice, D. (1996). Process modeling of the mechanics of mechanical alloying. *Scripta Mater.* **34**: 5–11.

32. Jurczyk, M., Jankowska, E., Nowak, M., and Jakubowicz, J. (2002). Nanocrystalline titanium type metal hydrides prepared by mechanical alloying. *J. Alloys Comp.* **336**: 265–269.
33. Jurczyk, M., Jankowska, E., Nowak, M., and Wieczorek, I. (2003). Electrode characteristics of nanocrystalline TiFe-type alloys. *J. Alloys Comp.* **354**: L1–L4.
34. Zaluski, L., Zaluska, A., and Ström-Olsen, J. O. (1995). Hydrogen absorption in nanocrystalline Mg$_2$Ni formed by mechanical alloying. *J. Alloys Comp.* **217**: 245–249.
35. Balcerzak, M., Jakubowicz, J., Kachlicki, T., and Jurczyk, M. (2015). Effect of multi-walled carbon nanotubes and palladium addition on the microstructural and electrochemical properties of the nanocrystalline Ti$_2$Ni alloy. *Int. J. Hydrogen Energy*, **40**: 3288–3299.
36. Li, X. D., Elkedim, O., Nowak, M., Jurczyk, M., Chassagnon, R. (2013). Structural characterization and electrochemical hydrogen storage properties of Ti$_{2-x}$Zr$_x$Ni (x = 0, 0.1, 0.2) alloys prepared by mechanical alloying, *Int. J. Hydrogen Energy* **38**: 12126–12132.
37. Zaluska, A., Zaluski, L., and Ström-Olsen, J. O. (2001). Structure, catalysis and atomic reactions on the nano-scale: A systematic approach to metal hydrides for hydrogen storage. *Appl. Phys. A* **72**: 157–165.
38. Edelstein, A. S., and Cammarata, R. R. (1996). *Nanomaterials: Synthesis, Properties and Applications*, Bristol (UK), Philadelphia (USA), Institute of Physics Publishing.
39. Iijima, S. (1991). Helical microtubules of graphite carbon. *Nature* **354**: 56–58.
40. Ajayan, P. M., Ebbesen, T. W., Ichihashi, T., Iijima, S., Tanigaki, K., and Hiura, H. (1993). Opening carbon nanotubes with oxygen and implications for filling. *Nature* **362**: 522–525.
41. Tsang, S. C., Chen, Y. K., Harris, P. J. F., and Green, M. L. H. (1994). A simple chemical method of opening and filling carbon nanotubes. *Nature* **372**: 159–162.
42. Peigney, A., Laurent, Ch., Flahaut, E., Bacsa, R. R., and Rousset, A. (2001). Specific surface area of carbon nanotubes and bundles of carbon nanotubes. *Carbon* **39**: 507–514.
43. Tibbets, G. G., Meisner, G. P., Olk, C. H. (2001). Hydrogen storage capacity of carbon nanotubes, filaments, and vapor-grown fibers. *Carbon* **39**: 2291–2301.
44. Cao, A., Zhu, H., Zhang, X., Li, X., Ruan, D., Xu, C., Wei, B., Liang, J., and Wu, D. (2001). Hydrogen storage of dense-aligned carbon nanotubes. *Chem. Phys. Lett.* **342**: 510–514.

45. Chen, C., and Huang, C. (2007). Hydrogen storage by KOH-modified multi-walled carbon nanotubes. *Int. J. Hydrogen Energy* **32**: 237–246.
46. Bogdanovic, B., and Schwickardi, M. (1997). Ti-doped alkali metal aluminum hydrides as potential novel reversible hydrogen storage materials. *J. Alloys Comp.* **253–254**: 1–9.
47. Kumar, R. S., Kim, E., and Cornelius, A. L. (2008). Structural phase transitions in the potential hydrogen storage compound KBH_4 under compression. *J. Phys. Chem. C* **112**: 8452–8457.
48. Liu, B. H., and Li, Z. P. (2009). A review: Hydrogen generation from borohydride hydrolysis reaction. *J. Power Sources* **187**: 527–534.
49. Shang, Y., and Chen, R. (2006). Hydrogen storage via the hydrolysis of $NaBH_4$ basic solution: Optimization of $NaBH_4$ concentration. *Energy Fuels* **20**: 2142–2148.
50. Züttel, A., Borgschulte, A., and Schapbach, L. (2008). *Hydrogen as a Future Energy Carrier*, Wiley.
51. King, R. B. (1994). *Encyclopedia of Inorganic Chemistry*, John Wiley & Sons, Chichester, UK.
52. http://www1.eere.energy.gov/hydrogenandfuelcells/storage/pdfs/targets_onboard_hydro_storage_explanation.pdf.
53. Kunowsky, M., Marco-Lózar, J. P., and Linares-Solano, A. (2013). Material demands for storage technologies in a hydrogen economy. Hindawi Publishing Corporation, *J. Renew. Energy* **2013**: Article ID 878329, 16 pages, http://dx.doi.org/10.1155/2013/878329.
54. Schlapbach, L., and Züttel, A. (2001). Hydrogen-storage materials for mobile applications. *Nature* (London) **414**: 353–358.
55. Bouten, P. C., and Miedema, A. R. (1980). On the heats of formation of the binary hydrides of transition metals. *J. Less-Common Met.* **71**: 147–160.
56. Buschow, K. H. J., Bouten, P. C. P., and Miedema, A. R. (1982). Hydrides formed from intermetallic compounds of two transition metals: A special class of ternary alloys. *Rep. Prog. Phys.*, **45**: 937–1039.
57. Szajek, A., Jurczyk, M., and Rajewski, W. (2000). The electronic structure and electrochemical properties of the $LaNi_5$, $LaNi_4Al$ and $LaNi_3AlCo$ systems. *J. Alloys Comp.* **307**: 290–296.
58. Szajek, A., Jurczyk, M., and Rajewski, W. (2000). The electronic and electrochemical properties of the ZrV_2 and $Zr(V_{0.75}Ni_{0.25})_2$ systems. *J. Alloys Comp.* **302**: 299–303.
59. Szajek, A., Jurczyk, M., Nowak, M., and Makowiecka, M. (2003). The electronic and electrochemical properties of the $LaNi_5$-based alloys. *Phys. Stat. Sol.* (a) **196**: 252–255.

60. Jurczyk, M., Nowak, M., Smardz, L., and Szajek, A. (2011). Mg-based nanocomposites for room temperature hydrogen storage, *TMS Annual Meeting* **1**: 229–236.
61. Wu, T. D., Xue, X. G., Zhang, T. B., Hu, R., Kou, H. C., and Li, J. H. (2016). Effect of MWCNTs on hydrogen storage properties of a Zr-based Laves phase alloy. *Int. J. Hydrogen Energy*, available online 1 February 2016; doi:10.1016/j.ijhydene.2015.12.114.
62. Jat, R. A., Parida, S. C., Agarwal, R., and Kulkarni, S. G. (2013). Effect of Ni content on the hydrogen storage behavior of $ZrCo_{1-x}Ni_x$ alloys. *Int. J. Hydrogen Energy* **38**: 1490–1500.
63. Hou, X. J., Hu, R., Zhang, T. B., Kou, H. C., and Li, J. S. (2013). Hydrogenation behavior of high-energy ball milled amorphous Mg_2Ni catalyzed by multi-walled carbon nanotubes. *Int. J. Hydrogen Energy* **38**: 16168–16176.
64. Ranjbar, A., Ismail, M., Guo, Z. P., Yu, X. B., and Liu, H. K. (2010). Effects of CNTs on the hydrogen storage properties of MgH_2 and MgH_2-BCC composite. *Int. J. Hydrogen Energy* **35**: 7821–7826.
65. Anani, A., Visintin, A., Petrov, K., and Srinivasan, S. (1994). Alloys for hydrogen storage in nickel/hydrogen and nickel/metal hydride batteries. *J. Power Sources* **47**: 261–275.
66. Kohno, T., Yoshida, H., Kawashima, F., Inaba, T., Sakai, I., Yamamoto, M., et al. (2000). Hydrogen storage properties of new ternary system alloys: La_2MgNi_9, $La_5Mg_2Ni_{23}$, La_3MgNi_{14}. *J. Alloys Comp.* **311**: L5–L7.
67. Kohno, T., Yoshida, H., Kawashima, F., Inabat, T., Sakai, I., Yamamoto, M., and Kanda, M. (2000). Hydrogen storage properties of new ternary system alloys: La_2MgNi_9, $La_5Mg_2Ni_{23}$, La_3MgNi_{14}, *J. Alloys Comp.* **311** (2000) L5–L7.
68. Zhang, F. L., Luo, Y. C., Chen, J. P., Yan, R. X., and Chen, J. H. (2007). La–Mg–Ni ternary hydrogen storage alloys with Ce_2Ni_7-type and Gd_2Co_7-type structure as negative electrodes for Ni/MH batteries. *J. Alloys Comp.* **430**: 302–307.
69. Chai, Y. J., Sakaki, K., Asano, K., Enoki, H., Akiba, E., and Kohno, T. (2007). Crystal structure and hydrogen storage properties of La-Mg-Ni-Co alloy with superstructure. *Scr. Mater.* **57**: 545–548.
70. Zhang, F. L., Luo, Y. C., Chen, J. P., Yan, R. X., Kang, L., and Chen, J. H. (2005). Effect of annealing treatment on structure and electrochemical properties of $La_{0.67}Mg_{0.33}Ni_{2.5}Co_{0.5}$ alloy electrodes. *J. Power Sources* **150**: 247–254.

Chapter 2

Nanomaterials

Maciej Tulinski and Mieczyslaw Jurczyk

*Institute of Materials Science and Engineering,
Poznan University of Technology, Poznan, Poland*

maciej.tulinski@put.poznan.pl, mieczyslaw.jurczyk@put.poznan.pl

2.1 Introduction

Nanotechnology is the science and technology of precise manipulation of the structure of matter at the nanoscale. A nanometer is a millionth of a millimeter or 10^{-9} meters. To be classified as a nanomaterial, the material must be less than 100 nm in size in at least one direction [1]. A nano-object is a material with at least one, two, or three dimensions in the nanoscale range of 1 to 100 nm and a nanoparticle is a nano-object with all three dimensions in the 1 to 100 nm range.

The limit of the size of nanomaterials to even 50 nm [2] or 100 nm [1] is justified by the fact that some physical properties of nanoparticles approach those of bulk when their size reaches these values. There is also a legitimate definition that extends this upper size limit to 1 micron, the sub-micron range being classified as nano.

Handbook of Nanomaterials for Hydrogen Storage
Edited by Mieczyslaw Jurczyk
Copyright © 2018 Pan Stanford Publishing Pte. Ltd.
ISBN 978-981-4745-66-6 (Hardcover), 978-1-315-36444-5 (eBook)
www.panstanford.com

Nanotechnology embraces many different fields and specialties, including engineering, physics, chemistry, electronics, and medicine, among others. Processes and functionality take place at the nanoscale and the resulting nanomaterial exhibits properties exceeding and sometimes even not available in the conventional materials.

2.2 Different Approaches to Produce Nanomaterials

There are many different ways to produce, i.e., nanomaterials (Table 2.1, Fig. 2.1) [3–5]. The traditional and simpler way of producing fine particles has been "top-down," referring to the reduction through attrition and various methods of comminution in the traditional sense. Nowadays, methods of production using "bottom-up" techniques are being increasingly utilized [6].

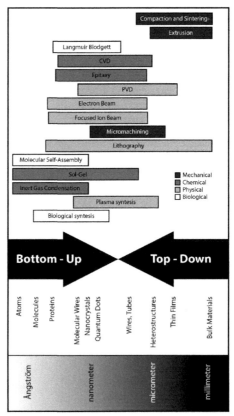

Figure 2.1 Schematic representation of the building up of nanostructures.

Table 2.1 A few of bottom-up and top-down methods

Top down	Bottom up
Mechanical alloying	Inert gas condensation
Etching	Sol-gel
Severe plastic deformation	Electron (or ion) beams
Lithography	Molecular self-assembly
High-energy ball milling	Ultrasonic dispersion
Micromachining methods	Metal-organic chemical vapor deposition (MOCVD)
Reactive milling	
	Electrodeposition
	Vacuum arc deposition
	Chemical/physical vapor deposition
	Molecular beam epitaxy

The bottom-up approach refers to the build-up of a material from the bottom: atom by atom, molecule by molecule, or cluster by cluster.

Both approaches play a very important role in modern industry and most likely in nanotechnology as well [5]. There are pros and cons of both approaches. The imperfection of surface structure and significant crystallographic damage to the processed patterns are the biggest problems with the top-down approach. However, despite the defects produced by the top-down approach, this approach will continue to play an important role in the synthesis of nanostructures.

An opposite approach is to synthesize the material from atomic or molecular species via chemical reactions, allowing for the precursor particles to grow in size. Currently, the bottom-up approach plays an important role in the fabrication and processing of nanostructures. It is important to note that the produced materials can have significantly different properties, depending on the route chosen to fabricate them. Apart from direct atom manipulation, there are various widely known methods for producing nanomaterials: physical, chemical, and mechanical (Table 2.2) [7, 8] and they can be done in gas, liquid, supercritical fluids, solid states, or vacuum.

Recently, the use of biological systems for the synthesis of nanoparticles has been proposed [9]. The nanoparticles synthesized by means of biogenic approach present good dispersity, dimensions, and stability.

Table 2.2   Synthesis methods of nanomaterials

Physical (bottom up)	Chemical (bottom up)	Mechanical (top down)
Physical vapor deposition	Chemical vapor deposition	Mechanical alloying
Laser vaporization	Epitaxial growth	High-energy ball milling
Laser pyrolysis	Colloidal dispersion	Mechanochemical synthesis
Ion beam techniques	Sol-gel	Mechanochemical activation synthesis
Nanolithography	Hydrothermal route	Severe plastic deformation
Plasma synthesis	Microemulsions	
Microwaves	Polymer route	

The biological methods of nanoparticle synthesis would allow the synthesis at physiological pH, temperature, pressure, and at the same time, at negligible cost. The physical and chemical methods are extremely pricey.

2.3   The Structure of Nanomaterials

The solids in nature have different internal order. The characteristic that distinguishes crystalline materials is the reproducibility of the atoms that are arranged in three main directions in space. The individual crystalline regions can be isolated from each other by non-crystalline layers with a thickness of 2 to 3 atoms. They are called the grain boundaries.

In contrast, the amorphous bodies have atoms placed completely randomly, and even within a few atoms adjacent to each other, one cannot extract the repeating sequence of atoms.

Nanomaterials are mostly polycrystalline solids, of which one of the characteristic dimensions is greater than 100 nm in at least one direction. Usually it is the size of the particle but may also be the thickness of the layers produced or applied to the substrate. In the case of three-dimensional nanomaterials, the term "grain" is not sufficiently precise, and therefore the term "crystallite" is used. Crystallite is the area that is coherently scattering X-rays, which is derived from a high degree of order within the crystallite material. A few crystallites combined in porousless assembly are called aggregate, in contrast to the agglomerate, which is porous and is composed of many crystallites.

The strength of the agglomerate (P_c) is described by Rumpf's equation [10]:

$$P_c = 9/8 \cdot (1 - V_p)/V_p \cdot F_k \, d^{-2},$$

where V_p is the porosity of the agglomerate, d the grain size, and F_k the mean tensile strength.

Nanocrystals can be pure metals, their alloys, and ceramic materials. The limit of size of nanocrystals is different for materials with different properties. Decreasing the size of nanocrystals generally involves the emergence of new quality of properties.

For example, in a nanocomposite $Nd_2Fe_{14}B/\alpha$-Fe-type permanent magnet, there is a phenomenon of strong interchangeable interaction between grains phases: the magnetically hard ($Nd_2Fe_{14}B$) and a magnetic soft (α-Fe), which results in the improvement of the remanence. This kind of phenomenon was observed in grains whose size was smaller than 30 nm [11]. Due to the large fraction of the volume of material occupied by grain boundaries, the phenomenon described above is greater in the solid materials than in thin layers. In the case of nanocrystalline materials, the volume of grain boundaries is about 50% for the grains of 5 nm, 5% for 10 nm and 3% for 100 nm [12].

Given the shape of the grains of nanocrystalline material, they can be divided into three groups (Fig. 2.2):

(1) Pillar: The particles have the shape of columns having a diameter of nanometers (one-dimensional).
(2) Sandwich: Grains have a flat shape with a thickness of nanometers (two-dimensional).
(3) Equiaxed: The particles have a shape close to a sphere having a diameter of nanometers (three-dimensional).

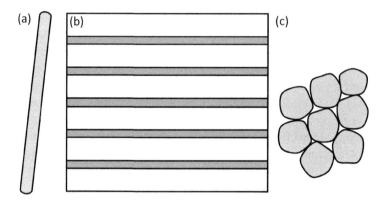

Figure 2.2 Nanocrystalline material types: (a) pillar, (b) sandwich, (c) equiaxed.

Given the chemical composition, nanomaterials can be divided into the following categories:

- the crystals and grain boundaries having the same chemical composition
- the crystals having a different chemical composition
- grains and a grain boundary phases having a different composition
- nanometric grains placed in the matrix of different composition

Nanocrystalline materials may consist of crystalline phases, crystalline and amorphous phases, and crystalline or amorphous matrix. An unfavorable feature of nanocrystals is their metastable character. The elevated temperature lowers the free energy of the system by reducing the energy of the grain boundaries. This results in grain growth.

As stated earlier, there are two general ways available to produce nanomaterials. The first way is to start with a bulk material and then break it into smaller pieces using mechanical, chemical, or other form of energy (top-down). An opposite approach is to synthesize the material starting from atomic or molecular scale via chemical reactions, allowing for the precursor particles to grow in size (bottom up). In the following sections, different methods of synthesizing nanomaterials in order to control particle size (and its distribution), particle shape and composition and degree of particle will be presented.

2.4 Methods of Synthesizing Nanomaterials

Hereafter, some different approaches to nanomaterials synthesis are briefly explained in the alphabetical order. Ball milling processes (i.e., mechanical alloying) are described in detail in Chapter 4.

2.4.1 Biological Synthesis

Nanoparticles can be synthesized using plant extracts, bacteria, fungi, yeast, and biological particles [9]. Biological synthesis of nanoparticles with controlled toxicity is rapidly growing and also it is an eco-friendly approach.

Biological particles like viruses, proteins, peptides, and enzymes could be exploited for biosynthesis of, i.e., Au/CdS and Au nanoparticles (see Fig. 2.3).

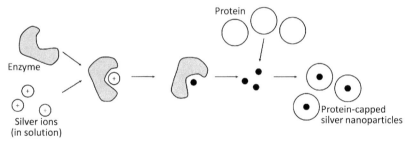

Figure 2.3 Schematic of biological synthesis of silver nanoparticles, using extracellular proteins.

The use of plants (e.g., *Geranium* extract, *Aloe vera* plant extracts, sun-dried *Cinnamomum camphora* and *Azadirachta indica* leaf extract) in the synthesis of Ag, Au nanoparticles provides single step biosynthesis process.

Ag nanoparticles can be also synthesized using bacteria (e.g., *Bacillus cereus, Bacillus thuringiensis, Escherichia coli, Lactobacillus, Pseudomonas stutzeri, Corynebacterium, Staphylococcus aureus, Ureibacillus thermosphaericus*). *Marinobacter pelagius* are also used to synthesize gold nanoparticles. Titanium nanoparticles were synthesized by the use of *Lactobacillus* strains.

Biological production of Ag nanoparticles by fungi (*Aspergillus niger, Aspergillus oryzae, Fusarium oxysporum, Fusarium solani,*

Pleurotus sajor-caju, Trichoderma viride) deals very well with low toxicity, higher bioaccumulation. It is also comparatively low-cost, effortless synthesis method with simple downstream processing and biomass handling.

2.4.2 Chemical Vapor Deposition

Chemical vapor deposition (CVD) is a processing technology that involves a chemical reaction between the substrate surface and a gaseous precursor [13]. The majority of its applications involve applying solid thin-film coatings to surfaces, but it is also used to produce high-purity bulk materials and powders, as well as fabricating composite materials. CVD is widely used to produce carbon nanotubes. The method itself involves flowing a precursor gas or gases into a chamber containing one or more heated objects to be coated (Fig. 2.4). Chemical reactions occur on and near the hot surfaces, resulting in the deposition of a thin film on the surface. This is accompanied by the production of chemical by-products that are exhausted out of the chamber along with unreacted precursor gases.

CVD films are generally quite conformal, materials can be deposited with very high purity and with relatively high deposition rates.

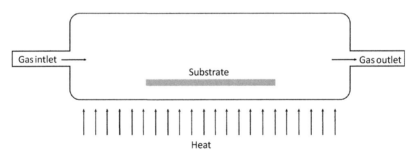

Figure 2.4 Schematic of a CVD deposition oven.

There are also disadvantages of chemical vapor deposition, mainly concerning the safety issue of the materials that are used for the deposition. It is known that some precursors and some by-products are toxic, pyrophoric, or corrosive.

2.4.3 Colloidal Dispersion

Colloidal dispersion is a heterogeneous system which is made up of dispersed phase and dispersion medium [14]. In colloidal dispersion, one substance is dispersed as very fine particles in another substance called dispersion medium (Fig. 2.5).

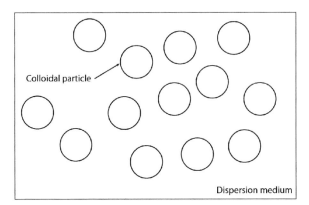

Figure 2.5 Colloidal dispersion system.

Dispersed phase and dispersion medium can be solid, liquid, or gaseous. Depending upon the state of dispersed phase and dispersion medium, colloidal dispersions can be divided into eight different types: foam, solid foam, liquid aerosol, emulsions, gels, solid aerosol, sol (colloidal suspension) and solid sol (solid suspension).

2.4.4 Epitaxial Growth

The ordered growth of single crystalline material on top of a substrate is called epitaxy [15]. Epitaxy is a non-equilibrium process where the driving force is proportional to the supersaturation, σ, which causes a transfer of material towards, and incorporation of material into the crystalline phase (Fig. 2.6). In a very simplified form, the equation for the flux J of material is given by: $J = k\sigma$, where k is the mass transport coefficient.

Some of the most common epitaxial techniques are liquid phase epitaxy (LPE), hydride vapor phase epitaxy (HVPE), metal organic vapor phase epitaxy (MOVPE), ultra high vacuum chemical vapor deposition (UHV-CVD), chemical beam epitaxy CBE, and molecular beam epitaxy (MBE).

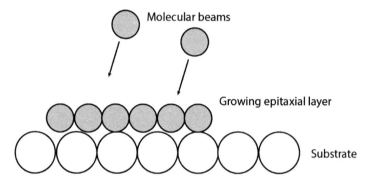

Figure 2.6 Rowing of epitaxial layer.

2.4.5 Hydrothermal Synthesis

Hydrothermal synthesis is a method of producing nanostructured materials with controlled size, shape distribution, and crystallinity (Fig. 2.7). Characteristics can be altered by changing experimental parameters, including reaction temperature, reaction time, solvent type, surfactant type, and precursor type [16].

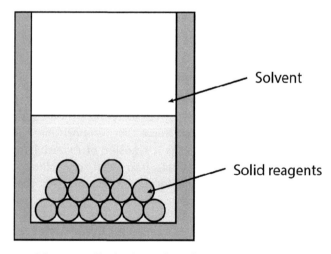

Figure 2.7 Schematic of hydrothermal synthesis.

This method has been used to synthesize nanostructured titanium dioxide, graphene, carbon, and other nanomaterials.

2.4.6 Ion Beam Techniques

Ion implantation involves the injection of very energetic ions into the surface of a solid substrate. The major components of an ion implantation facility are presented in Fig. 2.8 and include vacuum system, ion source, magnetic analyzer, accelerator, beam scanning, and target chamber.

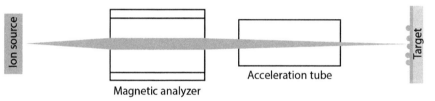

Figure 2.8 Schematic of ion implantation system.

To avoid scattering of the ions and contamination of the ion beam and target surface with unwanted ions, the entire ion implantation system should be equipped with ultra high vacuum (UHV) technology. Ions are generated within the ion source. Magnetic analyzer is used to obtain selectivity of the ions by accelerating them through a tube, where the ions are separated by charge to mass ratios. Then mass-selected ions are accelerated to high energies in acceleration tube and electrostatically focused into a spatially defined beam which is directed to the surface of a target.

High quenching rate of ion implantation techniques allows chemical and microstructural engineering at the atomic level. The main advantages of this technique are precision and flexibility, combined with the ability to obtain metastable crystalline and amorphous phases, non-equilibrium solid solutions, and nanocrystalline materials.

Examples of ion implantation include implantation of N to improve the wear resistance of metals, implantation of Ti and C into Fe-alloys to improve hardness and wear resistance.

Ion implantation's parameters can also be used to control nanocrystallite size distribution.

2.4.7 Lithography

The fabrication of nanoparticles has great significance for both fundamental scientific and technological applications [17]. Among all processes, lithography is the one that has been mainly used to create nanoparticles arrays (Fig. 2.9). The realization of these structures has depended largely on the development of nanolithography techniques, including photolithography, electron beam lithography, inkjet printing, fountain pen-based printing, soft lithography, and nanoimprinting. Unfortunately, these lithographic techniques can be expensive, complicated, and time consuming, especially when generating samples with large areas. In addition, the throughput and controllability of deposition processes of these nanoparticles remain poor.

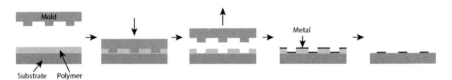

Figure 2.9 Illustration of the process of nanoimprint lithography (NIL).

Lithography involves the patterning of a surface through exposure to light, ions or electrons and then subsequent etching and/or deposition of material on that surface to produce the desired pattern. The ability to pattern features in the nanometer range is fundamental to the success of the IT industry. Electron- and ion-based methods are both capable of making sub-10 nm structures (with electron beam lithography having the greatest routine resolution), but they are too slow to be used directly in production. Optical lithography is used for production of semiconductor devices. Although it does not have the resolution of the beam-based techniques, it provides rapid throughput and cost-effective manufacturing. Electron beam lithography is primarily used to fabricate the masks used for optical lithography, while ion beam techniques are mostly used to repair masks.

2.4.8 Microemulsions

The microemulsion method is one of the most versatile preparation techniques which enables control of particle properties such as size, geometry, morphology, homogeneity, and surface area (Fig. 2.10) [18]. The microemulsion method has been used to synthesize colloidal metals, colloidal Fe_3O_4, colloidal AgCl, nanocrystalline Fe_2O_3, TiO_2, Al_2O_3, and high-Tc oxide $YBa_2Cu_3O_7$.

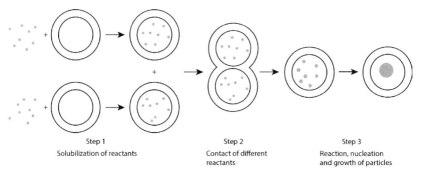

Figure 2.10 Microemulsion process.

Microemulsions are isotropic, macroscopically homogeneous, and thermodynamically stable solutions containing at least three components: a polar phase (usually water), a nonpolar phase (usually oil) and a surfactant. On a microscopic level, the surfactant molecules form an interfacial film separating the polar and the non-polar domains. This interfacial layer forms different microstructures ranging from droplets of oil dispersed in a continuous water phase (O/W-microemulsion) over a bicontinuous "sponge" phase to water droplets dispersed in a continuous oil phase (W/O-microemulsion).

2.4.9 Micromachining

Micromachining is a specific technique applied to micro- and meso-scale elements (Fig. 2.11) [19]. Along the rapid growth of micro electro mechanical systems (MEMS) research, the interest in manufacturing in microscopic scale is exponentially increasing. Although lithography-based manufacturing can achieve smaller feature size, micromachining has many advantages in terms of material choices, relative accuracy, and complexity of produced geometry. Moreover, it is a promising technology for bridging the gap between macro- and nano/micro domain.

Figure 2.11 Micromachining flowchart.

2.4.10 Physical Vapor Deposition

A vaporization coating technique involving transfer of material on an atomic level is commonly called physical vapor deposition (PVD). The raw materials/precursors start out in the solid form. PVD processes are carried out under vacuum conditions and involve following steps: evaporation, transportation, reaction, and deposition. During the first stage, a target (consisting of the material to be deposited) is bombarded by a high-energy source, such as a beam of electrons or ions. This "vaporates" atoms from the surface of the target. The next step simply consists of the transportation of "vaporized" atoms from the target to the substrate to be coated and generally takes place on straight trajectories.

The advantages of PVD coatings are improved hardness and wear resistance, reduced friction, and improved oxidation resistance. Thus, the main areas of application for PVD processes are thin films used in optical, optoelectronic, magnetic, and microelectronic devices. Other applications may be found in the areas of tribology, decorative coatings, corrosion protection, and thermal insulation.

One of the practical implementations of PVD technique is thermal evaporation (Fig. 2.12). For the evaporation process, the substance to be evaporated is heated (i.e., in a dedicated ceramic container, tungsten spiral wire, etc.) by the introduction of energy (in form of an electrical current, laser, electron beam, arc discharger etc.) to a suitable temperature. Also, to guarantee well defined film properties the substrate temperature often has to be as high as 100°C. The thermally released atoms or molecules leave the surface of the evaporated material and form a coating on the substrate or on the surrounding walls [8].

The evaporation technique uses thermal excitation to transfer the coating material into the gas phase, unlike sputtering techniques where the coating material is transferred into the gas phase by impulse transfer. This leads to the situation that basically each substance can be sputtered.

The basis of the widely spread sputter deposition process is the emission of neutral atoms, clusters, molecules or ions of the material bombarded by the energetic particles (Fig. 2.13).

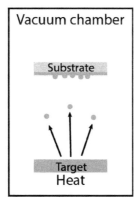

Figure 2.12 Schematic of evaporation system.

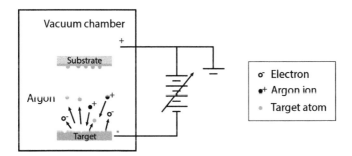

Figure 2.13 Schematic of sputtering system.

The positive ions generated from ionizing the sputtering gas by the glow discharge bombard the target surface and eject neutral atoms. The ejected atoms condense on the substrate surface and form into thin film.

The main limitations of sputtering techniques are difficulty in control of stoichiometry and low deposition rate and reproducibility, especially at the nanometer scale.

2.4.11 Plasma Synthesis

Low-temperature plasma is a versatile tool for synthesis, activation, and functionalization of various materials. In the nanosynthesis process, the material is first heated up to evaporation in induction plasma, and the vapors are subsequently subjected to a very rapid quenching in the quench zone (Fig. 2.14). The quench gas can be inert gases such as Ar and N_2 or reactive gases such as CH_4 and NH_3, depending on the type of nanopowders to be synthesized. The nanopowders produced are usually collected by porous filters, which are installed in a distance from the plasma reactor section. Because of the high reactivity of metal powders, special attention should be given to powder handling prior to the removal of the collected powder from the filtration section of the process.

The induction plasma system has been successfully used in the synthesis of nanopowders. The typical size range of the nanoparticles produced is from 20 to 100 nm, depending on the quench conditions employed. The productivity varies from a few hundred g/h to 3–4 kg/h.

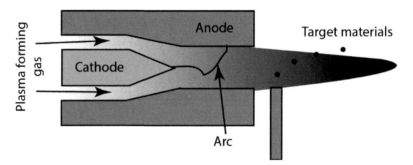

Figure 2.14 Schematic arrangement of plasma process.

In order to improve the surface properties of metals and expand their application, many methods have been applied to modify its surface. For example, properties of titanium were modified by boride microplasma surface alloying [20]. Plasma surface alloying gives a wide range of layer thickness, which is controlled by the amount of the powder and process parameters. Strong heat penetration from microplasma melt-in technique led to substrate dissolution with formation of stable TiB phase dispersed in α-Ti matrix. Plasma alloying is an effective method to produce TiB phase dispersed in α-Ti matrix with high hardness, good cytocompatibility, which makes them potential candidates for biomedical applications.

2.4.12 Polymer Route

Nanocomposites can be prepared by in situ synthesis of inorganic particles or by dispersion of fillers in a polymeric matrix (Fig. 2.15) [18]. A proper selection of the preparation technique is critical to obtain nanomaterials with suitable properties. The synthesis of polymer nanocomposites usually applies both bottom-up and top-down methodologies. In the bottom-up approach, precursors are used to construct and grow, from the nanometric level, into well-organized structures. Building block approaches can be used, too, where already formed entities or nano-objects are hierarchically combined to achieve the desirable material. The building block approach has an advantage compared to in situ nanoparticles formation, since at least one structural unit is well defined and usually does not have significant structural changes during the matrix formation.

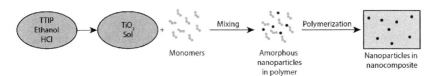

Figure 2.15 Schematic of in situ synthesis of nanoparticles in a polymer matrix.

2.4.13 Pulsed Laser Deposition

Pulsed laser deposition technique has been utilized to fabricate high-quality crystalline films, superior to films produced by other thin film growth techniques. PLD is one of the simplest and most versatile physical methods for deposition of thin films, including metals, semiconductors, nitrides, oxides, organic compounds and many more [21]. This technique is based on PVD (see Section 2.4.10) processes that are coming from the impact of high power short pulsed laser radiation on solid targets. As presented in Fig. 2.16, the laser beam enters the chamber and hits the target. The target and the substrate facing it are held in a vacuum chamber. The removal of atoms from the bulk material is done by the vaporization of the bulk at the surface region in a state of non-equilibrium. With PLD deposition of multilayer structures can be achieved, also film growth can be carried out in a reactive environment containing any kind of gas.

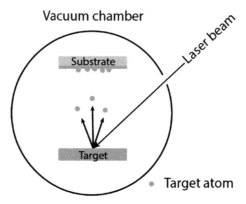

Figure 2.16 Schematic diagram of pulsed laser deposition.

There are many applications of PLD technique, for example, nanoscale metal oxide thin films, superconducting multilayered materials (YBCO) [22], nanocrystalline ZrC thin films with high wear resistance [23], etc. The deposition of ceramic oxides, nitride films, metallic multilayers and various superlattices has been also demonstrated.

2.4.14   Severe Plastic Deformation

Top-down approaches involve the refinement of coarse grains to ultra-fine grains. It can be achieved by severe plastic deformation (SPD) techniques [24]. With increasing straining of the material, the size of grains decreases in size and the disorientation is increasing. In order to obtain smallest microstructural sizes plastic strains more than 600% to 800% are necessary. A number of SPD methods for producing bulk ultra-fine grain metals/alloys have been developed [24, 25]. Currently, there is growing demand for the development of ultra-fine grain metallic materials for engineering applications because of their improved properties in comparison with bulk intermetallic alloys.

Severe plastic deformation techniques, such as equal channel angular processing (ECAP) [25], cyclic extrusion compression (CEC) [26], high pressure torsion (HPT) [27], twist extrusion (TE) [28], friction stir processing (FSP) [29], and multi-directional forging (MDF), also known as multiaxial compression/forging (MAC/F) [30], with the potential for intense straining, are employed to obtain ultra-fine grains (<500 nm), or nanocrystalline (<100 nm) materials. In addition, there are several methods of producing ultra-fine grain sheet metals, such as accumulative roll bonding (ARB) [31] and repeated corrugation and straightening (RCS) [32]. From different variants of SPD techniques, only a few have industrial potential due to the requirement of expensive and complicated die design and low productivity.

2.4.15 Sol-Gel

The sol-gel process involves the evolution of inorganic networks through the formation of a colloidal suspension (sol) and gelation of the sol to form a network in a continuous liquid phase (gel) (Fig. 2.17) [33]. The precursors for the synthesis of these colloids consist usually of a metal or metalloid element surrounded by various reactive ligands. The starting material is processed to form a dispersible oxide and forms a sol in contact with water or dilute acid. Removal of the liquid from the sol yields the gel, and the sol/gel transition controls the particle size and shape. Calcination of the gel produces the oxide.

Sol-gel processing refers to the hydrolysis and condensation of alkoxide-based precursors such as $Si(OEt)_4$. The reactions involved in the sol-gel chemistry based on the hydrolysis and condensation of metal alkoxides $M(OR)z$ can be described as follows:

$$MOR + H_2O \rightarrow MOH + ROH \text{ (hydrolysis)},$$

$$MOH + ROM \rightarrow M\text{-}O\text{-}M + ROH \text{ (condensation)}$$

The production of glasses by the sol-gel method permits the preparation of glasses at far lower temperatures than is possible by using conventional melting. It also makes possible synthesis of compositions that are difficult to obtain by conventional means due to the problems associated with volatilization, high melting temperatures, or crystallization. In addition, the sol-gel approach is a high-purity process that leads to excellent homogenity. The most successful applications utilize the composition control, microstructure control, purity, and uniformity of the method combined with the ability to form various shapes at low temperatures. Finally, the sol-gel approach can be adapted to producing films and fibers as well as bulk pieces.

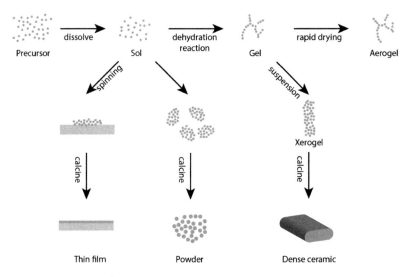

Figure 2.17 Sol-gel method.

Films and coatings were the first commercial applications of the sol-gel process. The development of sol-gel-based optical materials has also been quite successful, and applications include monoliths (lenses, prisms, and lasers), fibers (waveguides), and a wide variety of optical films. Other important applications of sol-gel technology utilize controlled porosity and high surface area for catalyst supports, porous membranes, and thermal insulation.

2.5 Summary

Nanotechnology is an interdisciplinary science established from chemistry, physics, and biology. By reducing the particle size, one can obtain unique physicochemical, biological, and mechanical properties. So far, different methods for the synthesis of nanoparticles in the field of nanotechnology have been

Table 2.3 Pros and cons of different synthesis methods

Method	Pros	Cons
Biological synthesis	–Controlled toxicity –Eco friendly	–Time consuming
CVD	–Uniform distribution –No compositional gradients	–High cost for compounds with sufficient purity –Safety and contamination issues
Epitaxial growth	–Good thickness control	–Growth rate and composition control limited by method
Hydrothermal route	–Good control of the size, shape and crystallinity	–High cost of equipment
Ion Beam	–Precision and flexibility –High quenching rate	–Ultra high vacuum needed
Lithography	–Ability to pattern features in the nanometer scale	–Expensive and time consuming
Microemulsions	–Good control of the size, geometry, morphology and homogenity	–Long-term stability of nanoparticles
Micromachining	–Accuracy and complexity of produced structures	–Time consuming
PVD	–Almost any material can be used –Environmentally friendly	–High vacuum and temperatures needed –Need of cooling equipment –Slow rate of deposition
Pulsed laser deposition	–Stoichiometric transfer of the target material –All kinds of materials can be deposited	–Ultra high vacuum needed –Possible stress-induced material breakdown
Severe plastic deformation	–High grain refinement –Fast process	–Low productivity –High cost of complicated die
Sol-gel	–High purity process –Good composition control	–High cost of the raw material –Volume shrinkage and cracking during drying

defined, each based on the reaction used, and have different advantages and disadvantages. Commercially, nanoparticles are mostly synthesized by chemical methods. In these methods, environmental pollution due to the material used in chemical synthesis are causes for the reduction of metal ions and stabilizing agents to prevent the accumulation of nanoparticles that are used in various processes.

While nanotechnology is seen as the way of the future and is a technology that a lot of people believe will bring a lot of benefit for all who will use it, nothing is ever perfect and there will always be pros and cons in each case (Table 2.3).

References

1. Borm, P. J. A., Robbins, D., Haubold, S., Kuhlbusch, T., Fissan, H., Donaldson, K., Schins, R. P. F., Stone, V., Kreyling, W., Lademann, J., Krutmann, J., Warheit, D., and Oberdorster, E. (2006). The potential risks of nanomaterials: A review carried out for ECETOC (review), *Part. Fibre Toxicol.* **3**: 1–35.

2. Kittelson, D. B. (2001) Recent measurements of nanoparticle emission from engines *Current Research on Diesel Exhaust Particles*, Japan Association of Aerosol Science and Technology, 9 January (Tokyo, Japan).

3. Dowling, A. P. (2004). Development of nanotechnologies, *Mater. Today*, 7: 30–35.

4. Bleeker, R. A., Troilo, L. M., and Ciminello, D. P. (2004). Patenting nanotechnology, *Mater. Today*, **2**: 44–48.

5. Zweck, A., Bachmann, G., Luther, W., and Ploetz, C. (2008). Nanotechnology in Germany: From forecasting to technological assessment to sustainability studies, *J. Cleaner Prod.* **16**: 977–987.

6. Scheu, M., Veefkind, V., Verbandt, Y., Galan, M. G., Absalom, R., and Förster, W. (2006). Mapping nanotechnology patents: The EPO approach, *World Patent Inf.* **28**: 204–211.

7. Wang, Z. L., Liu, Y., and Zhang, Z., eds. (2003). *Handbook of Nanophase and Nanostructured Materials,* Kluwer Academic/Plenum Publishers, New York, Chapter 1.

8. Yeadon, M., Yang, J. C., Ghaly, M., Olynick, D. L., Averbach, R. S., and Gibson, J. M. (1997). *Nanophase and Nanocomposite Materials 11*, Komarneni, S., Parker, J. C., and Wollenberger, H. J., eds., MRS, Pittsburgh, PA, pp. 179–184.

9. Ingale, A. G., and Chaudhari, A. N. (2013). Biogenic synthesis of nanoparticles and potential applications: An eco-friendly approach, *J. Nanomed. Nanotechol.* **4**: 165–172.
10. Rumpf, H. (1962). The strength of granules and agglomerates. In Knepper, W. A., ed., *Agglomeration*, Wiley-Interscience, pp. 379–418.
11. Jurczyk, M. (2000). Nanocomposite Nd-Fe-B-type magnets. *J. Alloys Comp.* **299**: 283–286.
12. Gleiter, H. (1990). Nanocrystalline materials, *Prog. Mater. Sci.* **33**: 223–315.
13. Craciun, D., Socol, G., Lambers, E., McCumiskey, E. J., Taylor, C. R., Martin, C., Argibay, N., Tanner, D. B., and Craciun, V. (2015). Optical and mechanical properties of nanocrystalline ZrC thin films grown by pulsed laser deposition, *Appl. Surface Sci.* **352**: 28–32.
14. Foord, J. S., Davies, G. J., and Tsang, W. T. (1997). *Chemical Beam Epitaxy and Related Techniques*, John Wiley & Sons Ltd, Chichester.
15. Dobkin, D., and Zuraw, M. K. (2003). *Principles of Chemical Vapor Deposition.* Kluwer.
16. Hayashi, H. and Hakuta, Y. (2010). Hydrothermal Synthesis of Metal Oxide Nanoparticles in Supercritical Water, *Materials* **3**: 3794–3817.
17. Huang, J. Y., Zhu, Y. T., Jiang, H., and Lowe, T. C. (2001). Microstructure and dislocation configuration in nanostructured Cu processed by repetitive corrugation and straightening, *Acta Mater.* **49**: 1497–1505.
18. Eastoe, J., and Warne, M. (1996). Nanoparticles and polymer synthesis in microemulsions, *Curr. Opin. Colloid. Interf. Sci.*, **1**: 800–805.
19. Huang, H. L., Chen, J.-K., Houng, M. P. (2012). Using soft lithography to fabricate gold nanoparticle patterns for bottom-gate field effect transistors, *Thin Solid Films* **524**: 304–308.
20. Miklaszewski, A., Jurczyk, M. U., Jurczyk, K., and Jurczyk, M. (2011). Plasma surface modification of titanium by TiB precipitation for biomedical applications, *Sur. Coat. Technol.* **206**: 330–337.
21. Verhoeven, J. D., Downing, H. L., Chumbley, L. S., and Gibson, E. D. (1989). The resistivity and microstructure of heavily drawn CuNb alloys, *J. Appl. Phys.* **65**: 1293–1301.
22. Chrisey, D. B., and Hybler, G. K. (1994). Pulsed *Laser Deposition of Thin Films*, Wiley & Sons.
23. Zhang, H., Yang, J., Wang, S., Wu, Y., Lv, Q., and Li, S. (2014). Film thickness dependence of microstructure and superconductive property of PLD prepared YBCO layers, *Physica C* **499**: 54–56.

24. Varin, R. A., Chiu, Ch., and Wronski, Z. S. (2008). Mechano-chemical activation synthesis (MCAS) of disordered Mg(BH$_4$)$_2$ using NaBH$_4$, *J. Alloys Comp.* **462**: 201–208.
25. Love, T. C., and Valiev, R. Z., eds. (2000). *Investigations and Applications of Severe Plastic Deformation* (Kluwer Academia Pub.).
26. Valiev, R. Z., Islamgaliev, R. K., and Alexandrov, I. V. (2000). Bulk nanostructured materials from severe plastic deformation, *Prog. Mater. Sci.* **45**: 103–189.
27. Richert, J., and Richert, M. (1986). A new method for unlimited. Deformation of metals and alloys, *Aluminium* **62**: 604–607.
28. Kim, H. S. (2001). Finite element analysis of high pressure torsion processing, *J. Mater. Proc. Technol.* **113**: 617–621.
29. Beygelzimer, Y., Varukhin, V., Synkov, S., Sapronov, A., and Synkov, V. (1999). New techniques for accumulating large plastic deformations using hydroextrusion, Fizika i Tekhnika Vusokih Davlenii (High Pressure Physics and Technology, in Russian). **6**: 499–508.
30. Thomas, W. M., Edward, N. D., Needham, J. C., Murch, M. G., Temple-Smith, P., Dawes, A., and Christopher, J. (1991). Friction Stir Welding, GB Patent Application 9125978.8. (1995). US Patent 5460317.
31. Fuloria, D., Goel, S., Jayanthan, R., Srivastava, D., Dey, G. K., Saibaba, N. (2015). Mechanical properties and microstructural evolution of ultrafine grained zircaloy4processed through multiaxial forging at cryogenic temperature, *Trans. Nonferrous Meter. Soc. China* **25**: 2221–2229.
32. Tsuji, N., Saito, Y., Utsunomiya, H., and Tanigawa, S. (1999). Ultra-fine grained bulk steel produced by accumulative roll-bonding (ARB) process, *Scrip. Mater.* **40**: 795–800.
33. Schramm, L. (2008). *Dictionary of Nanotechnology, Colloid and Interface Science*, Wiley.

Chapter 3

Solid-State Hydrides

Marek Nowak and Mieczyslaw Jurczyk

Institute of Materials Science and Engineering,
Poznan University of Technology, Poznan, Poland

marek.nowak@put.poznan.pl, mieczyslaw.jurczyk@put.poznan.pl

3.1 Metal Hydrides

Metal hydrides (MH_x) have been in focus for a long time as one of several alternatives to store hydrogen in a hydrogen-based economy [1–14]. These materials, which reversibly absorb and desorb hydrogen at ambient temperature and pressure, so-called "hydrogen storage alloys," are regarded as important materials for solving energy and environmental issues. Two other alternatives are high-pressure gas containers and liquid hydrogen. In metal hydrides, one uses the fact that hydrogen uptake in certain metals is exothermic and furthermore allows saturation uptakes that are close to, or may even exceed, that of liquid hydrogen. Researchers hope to use these compounds as a basis for storing hydrogen at higher densities than in pressurized tanks.

Hydrogen reacts with many elements to form hydrides. The binary hydride, a reaction product, can be of different types:

- metal hydrides

Handbook of Nanomaterials for Hydrogen Storage
Edited by Mieczyslaw Jurczyk
Copyright © 2018 Pan Stanford Publishing Pte. Ltd.
ISBN 978-981-4745-66-6 (Hardcover), 978-1-315-36444-5 (eBook)
www.panstanford.com

- ionic hydrides (LiH)
- covalent molecular hydrides (CH$_4$)

Figure 3.1 illustrates all the binary hydrides which can be formed by a direct reaction of hydrogen with the metal. The densities of the hydrogen in most of these metallic hydrides are greater than that of liquid hydrogen (Table 3.1). Most metallic hydrides are formed with pure hydrogen at room temperature.

Table 3.1 Hydrogen densities of some metal hydrides compared with liquid hydrogen

Compounds	PdH$_{0.8}$	MgH$_2$	TiH$_2$	VH$_2$	Liquid hydrogen
Hydrogen density (atom × 10^{-22}/cm^3)	4.7	6.6	9.2	10.4	4.2

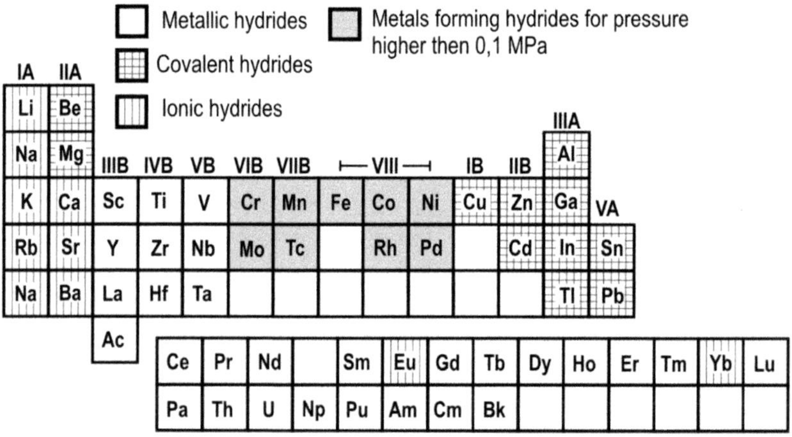

Figure 3.1 Metals forming binary hydrides.

3.2 Intermetallic Hydrides

A large number of the hydrogen storage materials have been studied until now. The first review on metal hydrides was written by Buschow et al. in 1982 [2]. Presently, a number of reviews are available on this topic [3–24]. Studies in metal-hydrogen grew rapidly in the 1970s. As the result, many hydride alloys (TiNi, TiFe, LaNi$_5$) were studied and developed. Early, in the 1980s, AB$_2$-type Ti/Zr–V–Ni–M systems and rare-earth-based AB$_5$-type Ln–Ni–Co–Mn–Al alloys were investigated. These systems obtained long life hydrogen storage electrode alloys, but the Ti/Zr system has a higher capacity. In the late 1980s and early 1990s, the non-stoichiometric AB_2 and AB_5 systems with improved performance such as higher capacity and long cycle life were developed. Some example compounds are $LaNi_5H_6$ and Mg_2NiH_4, which have hydrogen densities currently reported at 1–3.5%.

Now, there is a renewed interest in hydrogen storage materials, both conventional metal hydrides and new materials such as carbon nanostructures [25–30] and novel compounds [10, 13, 31, 32]. By making the metal storage material in the form of nanoparticles, two advantages can be gained, one related to the kinetics and the other to thermodynamics. The kinetics aspect is obvious; with nanoparticles, the uptake and release kinetics become much faster than with the more commonly used microcrystalline particles. The second aspect concerns the possibility to control the thermodynamic properties by tuning particle size. Since the thermodynamic properties of sufficiently small particles change due to, for instance, surface energy and elasticity/plasticity effects, the dissociation pressure may be varied for a given material by using the particle size as a tuning parameter.

In recent years, the interest in the study of nanostructured materials has been consistently increasing, stimulated by recent advances in material synthesis and characterization techniques and the realization that these materials exhibit many interesting and unexpected properties with a number of potential technological applications. For example, hydrogen storage nanomaterials are the key to the future of the storage and batteries/cells industries [9, 23, 24].

Recently, it has been shown that MA of TiFe, ZrV_2, $LaNi_5$, or Mg_2Ni is effective for the improvement of the initial hydrogen absorption rate, due to the reduction in the particle size and the creation of new clean surfaces [24, 33–36].

Independently, La–Mg–Ni compounds with A_2B_7 type structure are already utilized in novel Ni-MH_x batteries [37]. Due to low production costs and high discharge capacities (above 300 mAh/g) La–Mg–Ni-system A_2B_7-type alloys are considered to be the most promising candidates as the negative electrode materials of the Ni-MH_x rechargeable battery.

Proper engineering of microstructure by the use of unconventional processing techniques will lead to advanced nanocrystalline intermetallics representing a new generation of metal hydride materials [9, 33–36, 38–42].

The group of metallic hydrides is increased in number if, in addition to binary systems, one also includes hydrides based on intermetallic compounds of transition metals. A wide set of alloys composed of rare earth (RE) with nickel (AB_5-type) and alloys of zirconium and vanadium with nickel (AB_2-type), as well as titanium and magnesium with nickel (AB- or A_2B-type) was offered for use as hydrogen storage materials (Tables 3.2 and 3.3).

Table 3.2   Hydrides of binary intermetallic compounds

Hydrides	System
$TiFeH_2$	AB
$ZrV_2H_{5.5}$	AB_2
$LaNi_5H_6$	AB_5
$Mg_2NiH_{3.6}$	A_2B

Table 3.3   Hydride-forming alloys

Alloy and structure	Hydrides	Type of alloy
ZrV_2-cubic	$ZrV_2H_{5.5}$	AB_2
$LaNi_5$-hexagonal	$LaNi_5H_6$	AB_5
TiFe -cubic	$TiFeH_{1.7}$	AB
Mg_2Ni-hexagonal	Mg_2NiH_6 (at 350°C)	A_2B
$La_{0.8}Mg_{0.2}Ni_{3.75}$-hexagonal	$(La, Mg)_2Ni_7H$	A_2B_7

In all these alloys, component A is the one which forms the stable hydride. Component B performs several additional functions:

- It can play a catalytic role in enhancing the hydriding-dehydriding kinetic characteristics,
- It can alter the equilibrium pressure of the hydrogen absorption-desorption process to desired level,
- It can also increase the stability of the alloy, preventing the dissolution or formation of a compact oxide layer of component A.

The TiNi, ZrV$_2$, and, LaNi$_5$ phases are familiar materials which absorb large quantities of hydrogen under mild conditions of temperature and pressure. These types of hydrogen-forming compounds have recently proven to be very attractive as negative electrode material in rechargeable nickel-metal hydride batteries [23, 24, 40]. Magnesium-based hydrogen storage alloys have been also considered to be possible candidates for electrodes in Ni-MH batteries [9, 24, 41, 42].

The main example of the AB$_5$ class alloy is LaNi$_5$ with a hexagonal, CaCu$_5$-type structure containing three octahedral and three tetragonal sites per elemental cell unit [17, 20]. LaNi$_5$ forms two hydrides, one with low hydrogen content (α phase, LaNi$_5$H$_{03}$) and the other with high hydrogen content (β phase, LaNi$_5$H$_{5.5}$), which differ from each other significantly (about 25%) in the specific lattice volume. The discrete lattice expansion during conversion from the hydrogen solution phase (α) to the hydride phase (β) promotes crumbling of the alloy particles on hydriding-dehydriding cycles.

In the AB$_5$-type alloy, both La and Ni can be replaced by other elements (Mm, Ce, Pr, Nd, Zr, Hf→La and Al, Si, Cr, Mn, Fe, Co, Cu, Zn→Ni) [43–96]. Substitutions of A in AB$_5$ (La$_{1-y}$M$_y$B$_5$) by Zr, Ce, Pr, Nd decrease the unit cell volume and improve activation [43]. On the other hand, the use of Mm instead of La reduces the alloy costs.

Substitution of B in AB$_5$ (La(Ni$_{1-x}$M$_x$)$_5$ by M = Co, Cu, Fe, Mn, Al) by Ni is indispensable to prevent the decrease of the amount of absorbed hydrogen. Additionally, Co decreases the volume expansion upon hydriding, retards an increase of the internal cell pressure, but increases the alloy costs. Substitution of Co by Fe allows cost reduction without affecting cell performance, decreases decrepitation of alloy during hydriding. Al increases hydride formation energy. Mn decreases equilibrium pressure without decreasing the amount of stored hydrogen. V increases the lattice volume and enhances the hydrogen diffusion [44, 88, 91, 93].

Moreover, addition to B in AB$_5$ {(La,Mm)(Ni,M)$_{5-x}$B$_x$ M = Co, Cu, Fe, Mn, Al; B = A1, Si, Sn, Ge, In, Tl} was studied. The metals Al, Si, Sn, and Ge minimize the corrosion of the hydride electrode. Ge-substituted alloys exhibit better kinetics of hydrogen absorption/desorption in comparison with Sn-containing alloys.

In, Tl, and Ga prevent the generation of gaseous hydrogen [49, 50, 97, 98].

In non-stoichiometric $AB_{5\pm x}$ alloys (A = La; B = (Ni, Mn, Al, Co, V, Cu)), additional Ni forms separate finely dispersed phase. In $MmB_{5.12}$, the Ni_3Al-type second phase with high electrocatalytic activity is formed. Alloys poor in Mm are destabilized and the attractive interaction between the dissolved hydrogen atoms increases [23, 45, 46, 48, 57, 58, 60, 79, 89, 90, 99–103].

Furthermore, a partial replacement of the A and B components significantly changes the AB_5 alloy microstructure, e.g., Mn facilitates nucleation [21]. Also, as reported by Notten at al. [83, 104], the stoichiometry of an alloy influences its durability in the long-term hydriding-dehydriding cycles. Independently, it was found that the respective replacement of La and Ni in $LaNi_5$ by small amounts of Zr and Al resulted in a prominent increase in the cycle lifetime without causing much decrease in capacity [96]. A typical AB_5-type alloy consists of at least seven different metals, e.g., $La_{0.64}Ce_{0.36}Nd_{0.46}Ni_{0.95}Cr_{0.19}Mn_{0.41}Co_{0.15}$ (Hydralloy F) [105].

ZrV_2 and TiFe alloys crystallize in the cubic $MgCu_2$ and CsCl structures, and at room temperature they absorb up to 5.5 H/f.u. and 2 H/f.u., respectively. Nevertheless, the application of these types of materials in batteries has been limited due to slow absorption/desorption kinetics in addition to a complicated activation procedure. The properties of hydrogen host materials can be modified substantially by alloying to obtain the desired storage characteristics, e.g., proper capacity at a favorable hydrogen pressure. Significant improvements of the AB_2-type alloys have been achieved by additives contributing to the changes both in the unit cell volume and in the electronic structure, and, as a consequence, strongly affecting the stability and stoichiometry of the related hydrides [105, 106]. A serious problem in the Zr-containing alloy system is the formation of a compact ZrO_2 layer on the sample surface. Owing to the low electrical conductivity of this oxide, the transport of hydrogen into the catalytic sub-surface layer is impeded. This disadvantage in ZrV_2 can be eliminated by substitution, in which Zr is partially replaced by Ti and V is partially replaced by other transition metals (Cr, Mn and Ni) [105]. But the hydrogen storage capacity of ZrV_2 alloy decreases with increasing additive content.

For example, in ZrM$_2$-type alloy (M = V, Cr, Mn, Ni), an increase of the V content increases the maximum amount of absorbed hydrogen. Ni substitution decreases the electrochemical activity of an alloy. Composite alloy, mixture of ZrNi$_2$ with RNi$_5$ (RE = rare earth element), shows improved characteristics in comparison with the parent compounds.

In stoichiometric AB$_2$-type alloy, the distribution of A and B elements on the A and B sites is crucial for high hydrogen storage capacity; the over-stoichiometric alloy, in which some of the V atoms move from B to A sites, shows very high capacity.

Substitution of A in AB$_2$ by A = Zr + Ti:Zr contributes to an increase of the amount of stored hydrogen and induces formation of the C15-type desired phase structure (Zr) and increases the equilibrium pressure of hydrogen and decreases the electrochemical capacity (Ti). The amount of C15 phase decreases with increasing x in Ti$_x$Zr$_{1-x}$Ni$_{1.1}$V$_{0.5}$Mn$_{0.2}$Fe$_{0.2}$. At x = 0.5 it is pure C14 phase, and at x = 0.75 it is 13% cubic bcc phase + 87% C14 phase; the bcc phase absorbs more hydrogen than the C14 hexagonal phase. On the other hand, in substitution of B in AB$_2$ the Si, Mn-substituted Zr$_{0.8}$Ti$_{0.2}$(V$_{0.3}$Ni$_{0.6}$M$_{0.1}$)$_2$ alloys have C14 Laves phase structure, but the Co, Mo-substituted alloys form Cl5 Laves phase structure.

Additionally, Mn enhances the activation of an alloy during chemical pretreatment and increases discharge capacity. Co addition leads to the longest cyclic lifetime. Cr addition reduces the discharge capacity but extends cyclic lifetime; Cr controls the dissolution of V and Zr.

In non-stoichiometric alloys AB$_{2\pm x}$ (Zr$_{0.495}$Ti$_{0.505}$V$_{0.771}$Ni$_{1.546}$, Ti$_{0.8}$Zr$_{0.2}$Ni$_{0.6}$), increasing the Ni content decreases V-rich dendrite formations.

Typical AB$_2$-type alloys for hydrogen storage are multi-component and multi-phase systems. For example, a chemical composition of multi-component AB$_2$-type alloy is Zr$_{0.55}$Ti$_{0.45}$V$_{0.55}$Ni$_{0.88}$Cr$_{0.15}$Mn$_{0.24}$Co$_{0.18}$Fe$_{0.03}$ (patent Ovonic Battery Company) [23, 52, 53, 55, 61, 64–66, 68, 70–72, 76–78, 81]. The hydrogenation behavior of AB$_2$- and AB-type alloys is strongly influenced by their composition and stoichiometry, as well as by the microstructure of the final material and the presence of impurities [96, 105–118].

Among the different types of hydrogen-forming compounds, Ti-based alloys are among the promising materials for hydrogen energy applications [15, 107, 112–115, 117, 119–126]. For example, the TiFe alloy, which crystallizes in the cubic CsCl-type structure, is lighter and cheaper than the LaNi$_5$-type alloys and can absorb up to 2 H/f.u. at room temperature. It is a nontoxic material. Nevertheless, the application of TiFe material has been limited due to poor absorption/desorption kinetics in addition to a complicated activation procedure. To improve the activation of this alloy, several approaches have been adopted. For example, the replacement of Fe by some amount of transition metals to form secondary phase may improve the activation property of TiFe. After activation, TiFe reacts directly and reversibly with hydrogen to form two ternary hydrides TiFeH (orthorhombic) and TiFeH$_2$ (monoclinic), each of which is a distorted form of the bcc structure of the unhydrided alloy. In addition, excess Ti in TiFe, i.e., Ti$_{1+x}$Fe enables the alloy to be hydrided without the activation treatment. On the other hand, ball milling of TiFe is effective for the improvement of the initial hydrogen absorption rate, due to the reduction in the particle size and to the creation of new clean surfaces [127, 128]. Pd substitution in TiFe$_{1-x}$Pd$_x$ increases both the lattice constant and the catalytic activity, but decreases the plateau pressure.

Magnesium-based alloys have been extensively studied during last years, but the microcrystalline Mg$_2$Ni alloy can reversibly absorb and desorb hydrogen only at high temperatures [129, 130]. Substantial improvements in the hydriding-dehydriding properties of Mg-type metal hydrides could possibly be achieved by the formation of nanocrystalline structures [9, 131, 132]. It was found that the electrochemical activity of nanocrystalline hydrogen storage alloys can be improved in many ways, by alloying with other elements [24, 133], by ball-milling the alloy powders with a small amount of nickel or graphite powders [134–136]. For example, the surface modification of nanocrystalline hydrogen storage alloys with graphite by ball-milling leads to an improvement in both discharge capacity and charge-discharge cycle life. Recently, it has been demonstrated that the use of palladium and multi-wall carbon nanotubes (MWCNTs) offers many opportunities in tailoring hydrogen sorption properties of metal hydride [135, 136]. The catalytic effect of Pd and MWCNTs and MA process on Ti$_2$Ni alloys hydrogen storage properties have been studied in details.

3.3 Advanced Carbon Hydrides

Diamond and graphite are two forms of carbon. In diamond, each atom is fully coordinated symmetrically in space in all three dimensions and graphite is built up of a two-dimensional hexagonal sheet of carbon atoms, with long distance between each sheet. Additionally there are also other forms of carbon structures such as fullerenes, nanotubes, and graphene.

3.3.1 Fullerenes

Fullerenes are a new class of carbon aromatic compounds with unusual properties. Fullerenes are synthesized carbon molecules usually shaped like a football, such as C_{60} and C_{70}, and they are able to hydrogenate through the reaction:

$$C_{60} + xH_2O + xe- \circledcirc C_{60}H_x + xOH^- \qquad (3.1)$$

According to theoretical calculations, the most stable of these are $C_{60}H_{24}$, $C_{60}H_{36}$ and $C_{60}H_{48}$, latter of which is equal to 6.3 wt% of hydrogen adsorbed [137]. An experimental study by Chen et al. shows that more than 6 wt% of hydrogen can be adsorbed on fullerenes at 180°C and at about 25 bar. Despite the quite high hydrogen-storing ability, the cyclic tests of fullerenes have shown poor properties of storing hydrogen [137, 138].

3.3.2 Carbon Nanotubes

Nanotubes are the strongest carbon fibers that are currently known, single-wall nanotubes (SWNTs) and multi-wall nanotubes (MWNTs). Due to their large surface areas with relatively small mass, carbon nanotubes have been considered very promising potential materials for high-capacity hydrogen storage, as well [138–144]. The density functional calculations have shown that theoretically in proper conditions, a single-wall nanotube can adsorb over 14 wt% and a multi-wall nanotube about 7.7 wt% of hydrogen. Chen et al. reported that alkali-doped nanotubes are able to store even 20 wt% under ambient pressure but are unstable or require elevated temperatures [144].

Recent results on hydrogen uptake of single-wall nanotubes are promising. At 0.67 bar and 327°C, about 7 wt% of hydrogen have been adsorbed and desorbed with a good cycling stability. Another result at ambient temperature and pressure shows that 3.3 wt% hydrogen can be adsorbed and desorbed reproducibly and 4.2 wt% hydrogen with a slight heating.

3.3.3 Graphene

Hydrogen can basically be adsorbed on graphene in two different ways: either by physisorption, i.e., interacting by van der Waals forces, or by chemisorption, i.e., by forming a chemical bond with the C atoms. Recently, hydrogen sorption properties of graphene-related materials were studied by gravimetric and volumetric methods [145]. The H_2 uptake versus surface area trend revealed that hydrogen storage by graphene materials does not exceed 1 wt% at 120 Bar H_2 at ambient temperatures. A linear increase of hydrogen adsorption versus surface area was observed at liquid nitrogen temperature with maximal observed value of ~5 wt% for 2300 m^2/g sample. It can be concluded that bulk graphene-related materials do not exhibit superior hydrogen storage parameters compared to other nanostructured carbon materials like activated carbons, carbon nanotubes, or nanofibers.

3.4 Complex Hydrides

Sodium, lithium, and beryllium are the only elements lighter than magnesium that can also form solid-state compounds with hydrogen. The hydrogen content reaches 18 wt% for $LiBH_4$. The use of complex hydrides for hydrogen storage is challenging because of both kinetic and thermodynamic limitations [146].

Intense interest has developed in low-weight complex hydrides such as alanates $[AlH_4]^-$, amides $[NH_2]^-$, imides and borohydrides $[BH_4]^-$. In such systems, hydrogen is often located at the corners of a tetrahedron. The alanates and borates are especially interesting because of their light weight and the capacity for large number of hydrogen atoms per metal atom. Borates are known to be stable and decompose only at elevated temperatures. Alanates are remarkable due to their high storage capacities; however, they decompose in two steps upon dehydriding.

3.4.1 Alanates

Some of the lightest elements in the periodic table, for example, sodium, lithium, boron, and aluminum, form stable and ionic compounds with hydrogen (Table 3.4) [147]. The hydrogen content reaches values of up to 18 mass% for $LiBH_4$. However, such compounds desorb hydrogen only at temperatures from 80 to 600°C. The decomposition temperature of $NaAlH_4$ can be lowered by doping the hydride with TiO_2.

On the other hand, sodium aluminum hydride, $NaAlH_4$, could be a possible candidate for application as hydrogen storage material due to the theoretically reversible hydrogen storage capacity of 5.6 wt% and low cost [148, 149]. But, the complex aluminum hydrides are not considered as rechargeable hydrogen carriers due to irreversibility and poor kinetics. By using appropriate transition or rare-earth metals as catalysts, the complex hydrides can be made reversible. Upon doping with proper titanium compounds, the dehydriding of aluminum hydrides could be kinetically enhanced and maintain reversibility under moderate conditions in the solid state.

Lithium alanates are very attractive for hydrogen storage due of their high hydrogen content. The total hydrogen content is 10.5 and 11.2 wt% for $LiAlH_4$ and Li_3AlH_6, respectively. But $LiAlH_4$ composition is an unstable hydride, which decomposes easily and cannot be re-hydrogenated [13]. So, the commercialization of lithium-based compounds is hindered by their slow kinetics and high-temperature absorption and desorption.

Table 3.4 Crystal structure and hydrogenation/dehydrogenation properties of some alanates [13]

Material	Crystal structure	Hydrogen capacity (wt%)	Dehydrogenation temperature (°C)
$NaAlH_4$	Tetragonal	5.6	210–220 (1st step); >250 (2nd step)
$Mg(AlH_4)_2$	Trigonal	9.3	300 (1st step) 340 (2nd step)
Na_2LiAlH_6	Cubic	3.2	245

3.4.2 Nitrides and Other Systems

It was reported that lithium nitrides can store up 5.5 wt% of hydrogen after ball-milling of the LiNH$_2$ and LiH mixtures with the TiCl$_3$ as a catalyst [146]. Independently, Hu and Ruckenstein reported a reversible hydrogen capacity of 3.10 wt%, that increases with the number of cycles, due to improved porous structure of this material [150].

Table 3.5 Crystal structure and hydrogenation/dehydrogenation properties of some borohydrides [13]

Material	Crystal structure	Hydrogen capacity (wt%)	Dehydrogenation temperature (°C)	Dissociation enthalpy (kJ mol^{-1} H$_2$)
NaBH$_4$	cubic	10.8	400	−216.7 to −272.4
LiBH$_4$	orthorhombic	13.4	380	−177.0
Mg(BH$_4$)$_2$	—	13.7	260–400	−39.3 to −57.0
Ca(BH$_4$)$_2$	orthorhombic;	9.6	350	32

Boron is also an interesting element for hydrogen storage technologies (Table 3.5). LiBH$_4$ has a gravimetric hydrogen density of 18 wt% [151]. At temperatures greater than 470°C, hydrogen desorbs from LiBH$_4$. It was found that this compound releases hydrogen in different reaction steps and temperature regimes. The low-temperature desorption releases only 0.3 wt% of hydrogen and the high-temperature phase releases up to 13.5 wt%. LiBH$_4$ can reversibly store 8–10 wt% of hydrogen at temperatures of 315–400°C after the addition of MgH$_2$ with 2–3 mol% TiCl$_3$.

Besides alanates, nitrides and borohydrides, lithium–beryllium hydrides (Li$_3$Be$_2$H$_7$) are a new group of metal hydrides for hydrogen storage [152]. They show high reversible hydrogen capacity with more than 8 wt% at 150°C, but these materials are toxic.

References

1. Reilly, J. J., and Wismall, R. H. (1968). Metal hydrides for energy storage. *Inorg. Chem.* **7**: 2254–2256.

2. Buschow, K. H. J., Bouten, P. C. P., Miedema, A. R. (1982). Hydrides formed from intermetallic compounds from two transition metallic: A special class of ternary alloys. *Rep. Prog. Phys.* **45**: 937–1039.
3. Anani, A., Visintin, A., Petrov, K., Srinivasan, S. (1994). Alloys for hydrogen storage in nickel/hydrogen and nickel/metal hydride batteries. *J. Power Sources* **47**: 261–275.
4. Kopczyk, M., Wojcik, G., Młynarek, G., Sierczynska, A., and Beltowska-Brzezinska, M. (1996). Electrochemical absorption-desorption of hydrogen on multicomponent Zr-Ti-V-Ni-Cr-Fe alloys in alkaline solution. *J. Appl. Electrochem.* **26**: 639–645.
5. Grégoire Padró, C. E., Lau, F., eds. (2002). *Advances in Hydrogen Energy*, Kluwer Academic Publ.
6. Bououdina, M., Grant, D., and Walker, G. (2006). Review on hydrogen absorbing materials—structure, microstructure, and thermodynamic properties. *Int. J. Hydrogen Energy* **31**: 177–182.
7. Sakintuna, B., Lamari-Darkrim, F., and Hirscher, M. (2007). Metal hydride materials for solid hydrogen storage: A review, *Int. J. Hydrogen Energy* **32**: 1121–1140.
8. Jones, R. H., Thomas, G. J., eds. (2008). *Materials for the Hydrogen Economy*, CRC Press.
9. Varin, R. A., Czujko, T., and Wronski, Z. S. (2009). *Nanomaterials for Solid State Hydrogen Storage*, Springer Science+Business Media, LLC.
10. Jain, I. P. (2009). Hydrogen the fuel for 21st century. *Int. J. Hydrogen Energy*. **34**: 7368–7378.
11. Principi, G., Agresti, F., Maddalena, A., and Lo Russo, S. (2009). The problem of solid state hydrogen storage, *Energy* **34**: 2087–2091.
12. Zhao, X. G., and Ma, L. Q. (2009). Recent progress in hydrogen storage alloys for nickel/metal hydride secondary batteries—review. *Int. J. Hydrogen Energy* **34**: 1788–4796.
13. Jain, I. P., Jain, P., and Jain, A. (2010). Novel hydrogen storage materials: A review of lightweight complex hydrides. *J. Alloys Comp.* **503**: 303–339.
14. Young, K. H., and Nei, J. (2013). The current status of hydrogen storage alloy development for electrochemical applications. *Materials* **6**: 4574–4608.
15. Bradhurst, D. H. (1983). Metal hydrides for energy storage. *Metals Forum* **6**: 139–148.
16. Willems, J. G. (1984). Metal hydride electrodes. *Philips J. Res.* **39**: 1–94.

17. Schlapbach, L., ed. (1988). *Hydrogen in Intermetallic Compounds I. Electronic, Thermodynamic, and Crystallographic Properties, Preparation. Topics in Applied Physics*, vol. 63, Springer.
18. Iwakura, C., Fukuda, K., Senoh, H., Inoue, H., Matsuoka, M., and Yamamoto, Y. (1998). Electrochemical characterization of MmNi$_{4.0-x}$Mn$_{0.75}$Al$_{0.25}$Co$_x$ electrodes as a function of cobalt content. *Electrochim. Acta* **43**, 2041–2046.
19. Sandrock, G. (1999). A panoramic overview of hydrogen storage alloys from a gas reaction point of view. *J. Alloys Comp.* **293–295**: 877–888.
20. Schlapbach, L. (1992). In *Surface Properties and Activation, Hydrogen in Intermetallic Compounds* (Schlapbach, L., ed.), vol. II.
21. Sakai, T., Matsuoka, M., and Iwakura, C. (1995). In *Handbook on the Physics and Chemistry of Rare Earths* (Gschneidner, K. A., Eyring, L., eds), vol. 21. Elsevier, Amsterdam, pp. 135–180.
22. Dar, S. K., Ovshinsky, S. R., Gifford, P. R., Corrigan, D. A., Fetcenko, M. A., and Venkatesan, S. (1997). Nickel/metal hydride technology for consumer and electric vehicle batteries—a review and up-date. *J. Power Sources* **65**: 1–7.
23. Kleparis, J., Wojcik, G., Czerwinski, A., Skowronski, J., Kopczyk, M., and Beltowska-Brzezinska, M. (2001). Electrochemical behavior of metal hydrides. *J. Solid State Electrochem.* **5**: 229–249.
24. Jurczyk, M. (2004). The progress of nanocrystalline hydride electrode materials. *Bull. Pol. Acad. Sci. Tech. Sci.* **52**: 67–77.
25. Chambers, A., Park, C., Baker, R. T. K., and Rodriguez, N. M. (1988). Hydrogen storage in graphite nanofibers. *Phys. Chem.* B **102**: 4253–4256.
26. Enoki, T., Shindo, K., and Sakamoto, N. (1993). Electronic properties of alkali-metal-hydrogen-graphite intercalation compounds. *Z. Phys. Chem.* **181**: 75–82.
27. Stan, G., and Cole, M. W. (1998). Hydrogen adsorption in Nanotubes. *J. Low Temp. Phys.* **110**: 539–544.
28. Chen, P., Wu, X., Lin, J., and Tan, K. L. (1999). High H$_2$ uptake by alkali-doped carbon nanotubes under ambient pressure and moderate temperatures. *Science* **285**: 91–93.
29. Tanabe, Y., and Yasuda, E. (2000). Carbon alloys. *Carbon* **38**: 329–334.
30. Hirscher, M., Becher, M., Haluska, M., Quintel, A., Skalakova, V., Choi, Y. M., Dettlaff-Weglikowska, U., Roth, S., Stepanek, I., Bernier, P., et al. (2002). Hydrogen storage in carbon nanostructures, *J. Alloys Comp.* **330–332**: 654–658.

31. Vajo, J. J., Skeith, S., and Mertens, F. (2005). Reversible storage of hydrogen in destabilized LiBH₄. *J. Phys. Chem. B* **109**: 3719–3722.
32. Satyapal, S., Petrovic, J., Read, C., Thomas, G., and Ordaz, G. (2007). The US Department of Energy's National Hydrogen Storage Project: Progress towards meeting hydrogen powered vehicle requirements. *Catal. Today* **120**: 246–256.
33. Zaluska, A., Zaluski, L., Ström-Olsen, J. O. (2001). Structure, catalysis and atomic reactions on the nano-scale: A systematic approach to metal hydrides for hydrogen storage. *Appl. Phys. A* **72**: 157–165.
34. Jurczyk, M., Nowak, M. (2008). Nanomaterials for hydrogen storage synthesized by mechanical alloying. In: *Nanostructured Materials in Electrochemistry* (Ali, E., ed.), Wiley, Chapter 9.
35. Jurczyk, M., Smardz, L., Makowiecka, M., Jankowska, E., Smardz, K. (2004). The synthesis and properties of nanocrystalline electrode materials by mechanical alloying. *J. Phys. Chem. Sol.* **65**: 545–548.
36. Jurczyk, M., Nowak, M., Jankowska, E., and Gasiorowski, A. (2001). Nanocrystalline alloy hydride electrodes formed by mechanical alloying. *Mater. Sci. Eng.* **4**: 416–418.
37. Yartys, V. A., Riabov, A. B., Denys, R. V., Sato, M., and Delaplane, R. G. (2006). Novel intermetallic hydrides. *J. Alloys Comp.* **408**: 273–279.
38. Jurczyk, M., Smardz, L., Szajek, A. (2004). Nanocrystalline materials for NiMH batteries. *Mater. Sci. Eng. B* **108**: 67–75.
39. Zäch, M., Hägglund, C., Chakarov, D., Kasemo, B. (2006). Nanoscience and nanotechnology for advanced energy systems. *Curr. Opin. Solid State Mater. Sci.* **10**: 132–143.
40. Hong, K. (2001). The development of hydrogen storage alloys and the progress of nickel hydride batteries. *J. Alloys Comp.* **321**: 307–313.
41. Gasiorowski, A., Iwasieczko, W., Skoryna, D., Drulis, H., Jurczyk, M. (2004). Hydriding properties of nanocrystalline $Mg_{2-x}M_x$ Ni alloys synthesized by mechanical alloying (M = Mn, Al). *J. Alloys Comp.* **364**, 283–288.
42. Jurczyk, M., Okonska, I., Iwasieczko, W., Jankowska, E., Drulis, H. (2007). Thermodynamic and electrochemical properties of nanocrystalline Mg_2Cu-type hydrogen storage materials. *J. Alloys Comp.* **429**: 316–320.
43. Bridger, N. J., and Markin, T. L. (1976). GB Patent A-1,546,613.
44. Boter, P. A. (1977). US Patent 4,048,407.
45. Percheron-Guegan, A., Achard, J.-C., Bronoe, G., and Sarradin, J. (1977). Fr Patent A-2,382,774.

46. Percheron-Guegen, A., Achard, C. J., Loriers, J., Bonnemay, M., Bronoel, G., Sarradin, J., and Schlapbach, L. (1978). US Patent 4,107,405.
47. Osumi, Y., Suzuki, H., Kato, A., Oguro, K., and Nakane, M. (1983). US Patent 4,396,576.
48. Kanda, M., and Sato, Y. (1984). Eur Patent A-0 149 846 Al.
49. Willems, J. G. (1984). Metal hydride electrodes stability of LaNi$_5$-related compounds. *Philips. J. Res.* **39**, 1–94.
50. Willems, J. J. G., van Beek, J. R., and Buschow, K. H. (1984). Eur Patent A-0 142 878 Al.
51. Mohri, M. (1985). Eur Patent A-0 156 241 A2.
52. Sapru, K., Hong, K., Fetcenko, M., and Venkatesan, S. (1985). US Patent 4,551,400.
53. Sapru, K., Hong, K., Venkatesan, S., and Fetcenko, M. (1985). Eur Patent A-0 161 075 A2.
54. Percheron-Guegan, A., Achard, J. C., Bronoel, G., and Sarradin, J. (1986). US Patent 4,609,599.
55. Sapru, K., Reichman, B., Reger, A., and Ovshinsky, S. R. (1986). US Patent 4,623,597.
56. Yagasaki, E., Kanda, M., Mitsuyasu, K., and, Sato, Y. (1986). Eur Patent A-0 206 776 A2.
57. Heuts, J. J. F. G., and Willems, J. J. G. S. A. (1987). US Patent 4,699,856.
58. Heuts, J. J. F. G., and Frens, G. (1987). US Patent 4,702,978.
59. Heuts, J. J. F. G., and Willems, J. G. (1987). Eur Patent A-0 251 384 Al.
60. Ikoma, M., Kawano, H., Matsumoto, I., and Yanagihara, N. (1987). Eur Patent A-0 271 043.
61. Magnuson, G., Wolff, M., Lev, S., Jeffries, K., and Mapes, S. (1987). US Patent 4,670,214.
62. Miyake, J., and Kawamura, S. (1987). Efficiency of light energy conversion to hydrogen by the photosynthetic bacterium *Rhodobacter sphaeroides*. *Int. J. Hydrogen Energy* **39**: 147–149.
63. Percheron-Guegan, A., Achard, J.-C., Bronoe, G., and Sarradin, J. (1987). Fr Patent A-2,399,484.
64. Reichman, B., Venkatesan, S., Fetcenko, M. A., Jeffries, K., Stahl, S., and Bennett, C. (1987). Eur Patent A-0 273 625 A2.
65. Reichman, B., Venkatesan, S., Fetcenko, M. A., Jeffries, K., Stahl, S., and Bennett, C. (1987). US Patent 4,716,088.
66. Venkatesan, S., Reichman, B., and Fetcenko, M. A. (1987). Eur Patent A-0 273 624 A2.

67. Gamo, T., Moriwaki, Y., and Iwaki, T. (1988). Eur Patent A-0 293 660 A2.
68. Fetcenko, M. A. (1988). US Patent 5,002,730.
69. Mitsuyasu, K., Kanda, M., Takeno, K., and Kochiwa, K. (1988). Eur Patent A-0 284 063 Al.
70. Venkatesan, S., Reichman, B., and Fetcenko, M. A. (1988). US Patent 4,728,586.
71. Wolff, M., Fetcenko, M. A., Nuss, M. A., and Lijoi, A. L. (1988). Eur Patent A-0 347 507 Al.
72. Fetcenko, M. A. (1989). Eur Patent A-0 360 203.
73. Ikoma, M., Ito, Y., Yuasa, K., Matsumoto, I., and Hino, T. (1989). Eur Patent A-0 384 945 Al.
74. Ikoma, M., Kawano, H., Takahashi, O., Matsumoto, I., and Ikeyama, M. (1989). Eur Patent A-0 383 991 A2.
75. Ikoma, M., Kawano, H., Matsumoto, I., and Yanagihara, N. (1989). US Patent 4,837,119.
76. Fetcenko, M. A. (1990). Eur Patent A-0 410 935.
77. Fetcenko, M. A., Kaatz, T., Sumner, S. P., and LaRocca, J. (1990). US Patent 4,893,756.
78. Fetcenko, M. A., Sumner, S. A., and LaRocca, J. (1990). US Patent 4,948,423.
79. Gamo, T., Iwaki, T., and Shintani, A. (1990). Eur Patent A-0 413 029 Al.
80. Ikoma, M., Kawano, H., Ito, Y., Matsumoto, I. (1990). US Patent 4,935,318.
81. Wolf, M., Nuss, M. A., Fetcenko, M. A., Lijoi, A. L., Sumner, S. P., LaRocca, J., and Kaatz, T. (1990). US Patent 4,915,898.
82. Furukawa, N., Inoue, K., Nogami, M., Kameoka, S., and Tadokoro, M. (1991). US Patent 5,000,104.
83. Notten, P. H. L., and Hokkeling, P. (1991). Double phase hydride forming compounds: A new class of highly electrocatalytic materials. *J. Electrochem. Soc.* **138**: 1877–1882.
84. Bernd, F. (1992). Product information. Gesellschaft fur Elektrometallurgie, Numberg.
85. Nakamura, Y., Nakamura, H., Fujitani, S., and Yonezu, I. (1994). Homogenizing behaviour in a hydrogen-absorbing LaNi$_{4.55}$Al$_{0.45}$ alloy through annealing and rapid quenching. *J. Alloys Comp.* **210**: 299–230.

86. Chen, J., Dou, S. X., and Liu, H. K. (1996). Effect of partial substitution of La with Ce, Pr and Nd on the properties of LaNi$_5$-based alloy electrodes. *J. Power Sources* **63**: 267–227.

87. Adzie, G. D., Johnson, J. R., Mukerjee, S., Mcbreen, J., and Reilly, J. J. (1997). Function of cobalt in AB(5)H(x) electrodes. *J. Alloys Comp.* **253–254**: 579–582.

88. Cocciantelli, J. M., Bernard, P., Fernandez, S., and Atkin, J. (1997). The influence of Co and various additives on the performance of MmNi$_{4.3-x}$Mn$_{0.33}$Al$_{0.4}$Co$_x$ hydrogen storage alloys and Ni/MH prismatic sealed cells. *J. Alloys Comp.* **253–254**: 642–647.

89. Iwakura, C., Miyamoto, M., Inoue, H., Matsuoka, M., and Fukumoto, Y. (1997). Effect of stoichiometric ratio on discharge efficiency of hydrogen storage alloy electrodes. *J. Alloys Comp.* **259**: 129–134.

90. Aymard, L., Ichitsubo, M., Uchida, K., Sekreta, E., and Ikazaki. F. (1997). Preparation of Mg$_2$Ni base alloy by the combination of mechanical alloying and heat treatment at low temperature. *J. Alloys Comp.* **259**: L5–L8.

91. Zuttel, A., Chartouni, D., Gross, K., Spatz, P., Bachler, M., Lichtenberg, F., and Folzer, A. (1997). Relationship between composition, volume expansion and cyclic stability of AB$_5$-type metal hydride electrodes. *J. Alloys Comp.* **253**: 626–628.

92. Lichtenberg, F., Kohler, U., Kleinsorgen, K., Folzer, A., and Bouvier, A. (1998). US Patent 5,738,953.

93. Hu, W. K., Lee, H., Kim, D. M., Jeon, S. W., and Lee, J. Y. (1998). Electrochemical behaviors of low-Co Mm-based alloys as MH electrodes. *J. Alloys Comp.* **268**: 261–265.

94. Iwakura, C., Fukuda, K., Senoh, H., Inoue, H., Matsuoka, M., and Yamamoto, Y. (1998). Electrochemical characterization of MmNi$_{4.0-x}$Mn$_{0.75}$Al$_{0.25}$Co$_x$ electrodes as a function of cobalt content. *Electrochim. Acta* **43**: 2041–2046.

95. Saito, N., Takahashi, M., and Sasai, T. (1999). US Patent 5,916,519.

96. Pan, H., Ma, J., Wang, C., Chen, S., Wang, X., Chen, C., et al. (1999). Studies on the electrochemical properties of MlNi$_{4.3-x}$Co$_x$Al$_{0.7}$ hydride alloy electrodes. *J. Alloys Comp.* **293–295**: 648–652.

97. Adzie, G. D., Johnson, J. R., Mukerjee, S., Mcbreen, J., and Reilly, J. J. (1997). Function of cobalt in AB$_5$Hx electrodes. *J. Alloys Comp.* **253–254**: 579–582.

98. Bowman, R. C., Witham, C., Fultz, B., Ratnakumar, B. V., Ellis, T. W., and Anderson, I. E. (1997). Hydriding behavior of gas-atomized AB$_5$ alloys. *J. Alloys Comp.* **253**: 613–616.
99. Percheron-Guegan, A., Achard, J.-C., Bronoe, G., Sarradin, J. (1987). Fr Patent A-2,399,484.
100. Gamo, T., Moriwaki, Y., and Iwaki, T. (1988). Eur. Patent A-0 293 660 A2.
101. Ikoma, M., Ito, Y., Yuasa, K., Matsumoto, I., and Hino, T. (1989). Eur. Patent A-0 384 945 Al.
102. Ikoma, M., Kawano, H., Takahashi, O., Matsumoto, I., and Ikeyama, M. (1989). Eur. Patent A-0 383 991 A2.
103. Ikoma, M., Kawano, H., Matsumoto, I., and Yanagihara, N. (1989). US Patent 4,837,119.
104. Ovshinsky, S. R., Fetcenko, M. A., and Ross, J. (1993). A nickel metal hydride battery for electric vehicles. *Science* **260**: 176–181.
105. Majchrzycki, W., and Jurczyk, M. (2001). Electrode characteristics of nanocrystalline (Zr,Ti)(V,Cr,Ni)$_{2.41}$ compound. *J. Power Sources* **93**: 77–81.
106. Wojcik, G., Kopczyk, M., Młynarek, G., Majchrzycki, W., and Beltowska-Brzezinska, M. (1996). Electrochemical behaviour of multicomponent Zr-Ti-V-Mn-Cr-Ni alloys in alkaline solution. *J. Power Sources* **58**: 73–78.
107. Liu, J. (1978). US Patent 4,111,689.
108. Sandrock, G. D. (1978). US Patent 4,079,523.
109. Shaltiel, D., Davidov, D., and Jacob, I. (1979). US Patent 4,163,666.
110. van Mal, H. H., van Esveld, H. A., van Wieringen, J. S., and Buschow, K. H. J. (1981). US Patent 4,283,226.
111. de Pous, O. (1981). US Patent 4,278,466.
112. Liu, J., and Lundin, C. E. (1982). US Patent 4,358,316.
113. Bernauer, O., and Ziegler, K. (1984). US Patent 4,446,101.
114. Gondo, H., Matsumoto, R., Ohno, J., and Suzuki, R. (1984). US Patent 4,488,906.
115. Suzuki, R., Ohno, J., and Gondo, H. (1986). US Patent 4,576,639.
116. Gamo, T., Moriwaki, Y., Iwaki, T., and Shintani, A. (1994). US Patent 5,281,390.
117. Wakao, S. (1995). US Patent 5,401,463.

118. Aoki, Y., and Miyaki, T. (2004). US Patent 6,787,103.
119. Buschow, K. H. J., Bouten, P. C. P., and Miedema, A. R. (1982). Hydrides formed from intermetallic compounds of two transition metals: A special class of ternary alloys. *Rep. Prog. Phys.* **45**: 937–1039.
120. Smardz, K., Smardz, L., Jurczyk, M., and Jankowska, E. (2003). Electronic properties of nanocrystalline and polycrystalline TiFe$_{0.25}$Ni$_{0.75}$ alloys. *Phys. Stat. Sol.* (a) **196**: 263–266.
121. van Mal, H. H., van Esveld, H. A., van Wieringen, J. S., and Buschow, K. H. J. (1981). US Patent 4,283,226.
122. Doi, H., Yabuki, R. (1990). US Patent 4,898,794.
123. Wakao, S. (1995). US Patent 5,401,463.
124. Nakamura, Y., Nakamura, H., Kamikawa, M., Watanabe, H., Fujitani, S., and Yonezu, I. (1998). US Patent 5,851,690.
125. Zaluski, L., Zaluska, A., and Ström-Olsen, J. O. (1997). Nanocrystalline metal hydrides. *J. Alloys Comp.* **253–254**: 70–79.
126. Jurczyk, M., Jankowska, E., Nowak, M., Wieczorek, I. (2003). Electrode characteristics of nanocrystalline TiFe-type alloys. *J. Alloys Comp.* **354**: L1–L4.
127. Selvam, P., Viswanathan, B., Swamy, S., and Srinivasan, V. (1986). Magnesium and magnesium alloy hydrides. *Int. J. Hydrogen Energy* **11**: 169–192.
128. Jung, C. B., Kim, J. H., Lee, K. S. (1997). Electrode characteristics of nanostructured TiFe and ZrCr$_2$ type metal hydride prepared by mechanical alloying. *Nano Struct. Mater.* **8**: 1093–1104.
129. Zaluski, L., Zaluska, A., and Ström-Olsen, J. O. (1995). Hydrogen absorption in nanocrystalline Mg$_2$Ni formed by mechanical alloying. *J. Alloys Comp.* **217**: 245–249.
130. Gennari, F. C., Castro, F. J., and Andrade-Gamboa, J. J. (2002). Synthesis of Mg$_2$FeH$_6$ by reactive mechanical alloying: Formation and decomposition properties. *J. Alloys Comp.* **339**: 261–267.
131. Anani, A., Visintin, A., Petrov, K., Srinivasan, S. (1994). Alloys for hydrogen storage in nickel/hydrogen and nickel/metal hydride batteries. *J. Power Sources* **47**: 261–275.
132. Orimo, S. I., and Fujii, H. (1998). Effects of nanometer-scale structure on hydriding properties of Mg–Ni alloys: A review. *Intermetallics* **6**: 185–192.
133. Au, M., Pourarian, F., Simizu, S., Sankar, S. G., and Zhang, L. (1995). Electrochemical properties of TiMn$_2$-type alloys ball-milled with nickel powder. *J. Alloys Comp.* **223**: 1–5.

134. Bouaricha, S., Dodelet, J. P., Guay, D., Huot, J., and Schultz, R. (2001). Activation characteristics of graphite modified hydrogen absorbing materials. *J. Alloys Comp.* **325**: 245–251.

135. Balcerzak, M., Jakubowicz, J., Kachlicki, T., and Jurczyk, M. (2015). Effect of multi-walled carbon nanotubes and palladium addition on the microstructural and electrochemical properties of the nanocrystalline Ti$_2$Ni alloy. *Int. J. Hydrogen Energy* **40**: 3288–3299.

136. Balcerzak, M., Jakubowicz, J., Kachlicki, T., and Jurczyk, M. (2015). Hydrogenation properties of nanostructured Ti$_2$Ni-based alloys and nanocomposites. *J. Power Sources* **280**: 435–445.

137. David, E. (2005). An overview of advanced materials for hydrogen storage. *J. Mater. Proc. Tech.* **162–163**: 169–177.

138. Cao, A., Zhu, H., Zhang, X., Li, X., Ruan, D., Xu, C., Wei, B., Liang, J., and Wu, D. (2001). Hydrogen storage of dense-aligned carbon nanotubes. *Chem. Phys. Lett.* **342**: 510–514.

139. Iijima, S. (1991). Helical microtubules of graphite carbon. *Nature* **354**: 56–58.

140. Ajayan, P. M., Ebbesen, T. W., Ichihashi, T., Iijima, S., Tanigaki, K., and Hiura, H. (1993). Opening carbon nanotubes with oxygen and implications for filling. *Nature* **362**: 522–525.

141. Tsang, S. C., Chen, Y. K., Harris, P. J. F., and Green, M. L. H. (1994). A simple chemical method of opening and filling carbon nanotubes. *Nature* **372**: 159–162.

142. Peigney, A., Laurent, Ch., Flahaut, E., Bacsa, R. R., and Rousset, A. (2001). Specific surface area of carbon nanotubes and bundles of carbon nanotubes. *Carbon* **39**: 507–514.

143. Tibbets, G. G., Meisner, G. P., and Olk, C. H. (2001). Hydrogen storage capacity of carbon nanotubes, filaments, and vapor-grown fibers. *Carbon* **39**: 2291–2301.

144. Chen, C., and Huang, C. (2007). Hydrogen storage by KOH-modified multi-walled carbon nanotubes. *Int. J. Hydrogen Energy* **32**: 237–246.

145. Klechikov, A. G., Mercier, G., Merino, P., Blanco, S., Merino, C., and Talyzin, A. V. (2015). Hydrogen storage in bulk graphene-related materials. *Microporous Mesoporous Mater.* **210**: 46–51.

146. Sakintuna, B., Lamari-Darkrim, F., and Hirscher, M. (2007). Metal hydride materials for solid hydrogen storage: A review. *Int. J. Hydrogen Energy* **32**: 1121–1140.

147. Bogdanovic, B., Brand, R. A., Marjanovic, A., Schwickardi, M., Tölle, J. (2000). Metal-doped sodium aluminium hydrides as potential new hydrogen storage materials. *J. Alloys Comp.* **302**: 36–58.

148. Jensen, C. M., Zidan, R., Mariels, N., Hee, A., and Hagena, C. (1999). Advanced titanium doping of sodium aluminum hydride segue to a practical hydrogen storage material. *Int. J. Hydrogen Energy* **24**: 461–465.

149. Sandrock, G., Gross, K. J., and Thomas, G. (2002). Effect of Ti-catalyst content on the reversible hydrogen storage properties of the sodium alanates. *J. Alloys Comp.* **339**: 299–308.

150. Hu, Y. H., and Ruckenstein, E. (2006). Hydrogen storage of Li_2NH prepared by reacting Li with NH_3. *Ind. Eng. Chem. Res.* **45**: 182–186.

151. Fakioglu, E., Yürüm, Y., and Veziroglu, T. N. (2004). A review of hydrogen storage systems based on boron and its compounds. *Int. J. Hydrogen Energy* **29**: 1371–1376.

152. Zaluska, A., Zaluski, L., and Ström-Olsen, J. O. (2000). Lithium-beryllium hydrides: The lightest reversible metal hydrides. *J. Alloys Comp.* **307**: 57–166.

Chapter 4

Preparation Methods of Hydrogen Storage Materials and Nanomaterials

Marek Nowak and Mieczyslaw Jurczyk

Institute of Materials Science and Engineering,
Poznan University of Technology, Poznan, Poland

marek.nowak@put.poznan.pl, mieczysław.jurczyk@put.poznan.pl

4.1 Introduction

Conventionally, microcrystalline hydride materials have been prepared by arc or induction melting and annealing. However, either a low storage capacity by weight or poor absorption-desorption kinetics in addition to a complicated activation procedure have limited the practical use of metal hydrides. Substantial improvements in the hydriding-dehydriding properties of metal hydrides could be possibly achieved by the formation of nanocrystalline structures by non-equilibrium processing techniques such as mechanical alloying (MA) [1–18] or high-energy ball milling (HEBM) [2, 6, 10, 15, 19]. In MA, elemental powdered materials are used as the starting materials, while in HEBM the starting material is an alloy with a desired composition.

Several hydrogen storage systems are available currently to achieve the common goals of hydrogen storage materials; the

---

Handbook of Nanomaterials for Hydrogen Storage
Edited by Mieczyslaw Jurczyk
Copyright © 2018 Pan Stanford Publishing Pte. Ltd.
ISBN 978-981-4745-66-6 (Hardcover), 978-1-315-36444-5 (eBook)
www.panstanford.com

goals are (i) high density of hydrogen, (ii) fast kinetics and appropriate thermodynamic properties, (iii) long-term cycling stability and high degree of reversibility, and (iv) cost. The method for preparing hydrogen storage materials evolves with the development of alloying techniques. It also varies according to the type of hydrogen storage material. The properties of hydrogen storage materials have been significantly improving through chemistry and microstructure modifications.

A large number of experimental investigations on $LaNi_5$-, TiFe-, TiNi-, and ZrV_2-type compounds have been performed up to now in relation to their exceptional hydrogenation properties [20–25]. Hydrogen storage material research has entered a new and exciting period with the development of nanocrystalline alloys, which show substantially enhanced absorption/desorption kinetics.

Nanoparticles are mainly produced in two ways (Fig. 4.1) [4, 26–28]. A nanoparticle is a nano-object with all three dimensions in the 1 to 100 nm range and showing properties not evident in the bulk material. The traditional way of producing fine particles has been "top-down," which refers to the reduction through attrition and various methods of comminution in the traditional sense. In the past years, methods of production using "bottom-up" techniques are being increasingly utilized. The bottom-up approach refers to the buildup of a material from the bottom: atom by atom, molecule by molecule, or cluster by cluster (see Chapter 2). Both approaches play a very important role in modern industry and most likely in nanotechnology as well. There are advantages and disadvantages of both approaches. The biggest problem with the top-down approach is the imperfection of the surface structure and significant crystallographic damage to the processed patterns. Regardless of the defects produced by the top-down approach, it will continue to play an important role in the synthesis of nanostructures.

It is important to note that the produced materials can have significantly different properties, depending on the route chosen to fabricate them. Apart from direct atom manipulation, there are various widely known methods for producing nanomaterials: physical, chemical, and mechanical. Several hydrogen storage systems are available currently, such as metallic compounds, complex hydrides, nanomaterials (nanotubes, nanofibers, nanoparticles, nanohorns), metal organic frameworks, calthrate hydrides, molecular sieve, and so on, to achieve the common goals of hydrogen storage materials [22, 24].

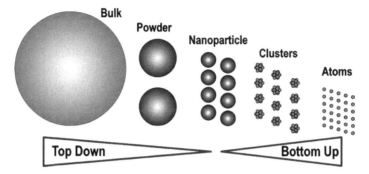

Figure 4.1 Schematic representation of the building up of nanostructures.

The physical and chemical methods are extremely expensive. In this chapter, we concentrate on nanoprocessing conducted entirely in a solid state (Table 4.1). This is a class of methods based on processes conducted in high-energy ball mills. Such methods are based on high-energy milling and grinding of microcrystalline materials. They provide the top-down approach toward manufacturing hydrogen storage nanomaterials.

Table 4.1 Pros and cons of "top-down" synthesis methods

Method	Pros	Cons
Mechanical alloying	–Low cost –High production rates of nanopowders	–Possible contamination by the milling tools
High energy ball milling	–Low cost –High production rates of nanopowders	–Possible contamination by the milling tools –Residual strain in the crystallized phase
Mechanochemical synthesis	–Low cost, easy process	–Possible contamination by the milling tools

In this chapter, we will introduce and discuss some of the most used methods for the preparation of hydrogen storage materials: melting and mechanical milling (mechanical alloying (MA) and high-energy bal-milling (HEBM)).

4.2 Microcrystalline Hydride Materials

Conventionally, the microcrystalline hydride materials have been prepared by arc or induction melting and annealing [29–31]. Melting is the most common method used on metallic compound system. Arc melting, induction melting, and melt spinning have all been applied on the preparation of hydrogen storage alloys. Arc melting (Fig. 4.2) and induction melting are conventional alloying methods. Arc melting is carried out in an electric arc furnace that heats charged material by means of an electric arc. It differs from induction melting in that the charge material is directly exposed to an electric arc, and the current in the furnace terminals passes through the charged material, while in induction melting process, the heat is applied by induction heating (Fig. 4.3) of melting.

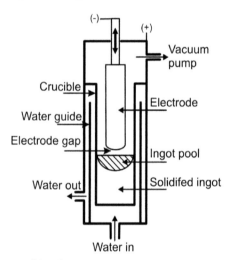

Figure 4.2 An arc melting furnace.

In the melt spinning (MS) method, alloy liquids were poured on a cooper or steel rotating wheel that is cooled internally by water or liquid nitrogen, and a thin stream of liquid alloy drips onto the wheel (Fig. 4.4). This cooling process makes rapid solidification [32]. The cooling rates is so high (10^4–10^7 K/s) that it can form metallic glasses. Flow turning of the melt occurs when a jet of molten alloy, under excessive inert gas pressure (argon) is fed through a hole in the quartz or ceramic crucible to the surface of a cooled drum rotating at a high speed. The resulting alloy's structure and properties depend on its composition, cooling rate and injection conditions.

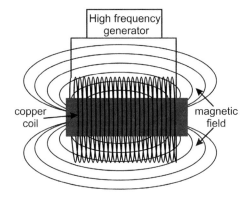

Figure 4.3 Induction heating.

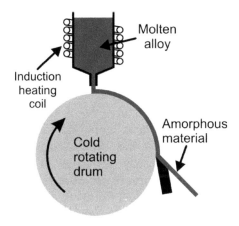

Figure 4.4 Melt spinning.

4.3 Nanotechnology for the Storage of Hydrogen

4.3.1 Mechanical Methods

Many different methods for producing small-grained nanostructures are available. They are mainly based on the production of fine-grained powders (down the nanometer scale). Mechanical processes include mechanical grinding, high-energy ball milling, mechanical alloying, and reactive milling [14–19]. The advantages of these techniques are that they are simple, require low-cost equipment, and, provided that a coarse feedstock powder can be made, the powder can be processed. However, there can be difficulties such as agglomeration of the powders and contamination from the process equipment. Mechanical processes are commonly used only for inorganics and metals/alloys.

In high-intensity ball milling, high-impact collisions are used to reduce microcrystalline materials down to nanocrystalline structures without chemical change. A relatively new technique termed mechanochemical processing (MCP) is a novel solid-state process for the manufacture of a wide range of nanopowders [33]. Dry milling is used to induce chemical reactions through ball/powder collisions that result in nanoparticles formed within a salt matrix. Particle size is defined by the chemistry of the reactant mix, milling, and heat treatment conditions. Particle agglomeration is minimized by the salt matrix, which is then removed by a simple washing procedure.

4.3.1.1 Milling processes

Mechanical alloying (MA) is a powder processing technique that allows production of homogeneous materials starting from blended elemental powder mixtures. MA is a dry, high-energy ball milling technique and has been employed to produce a variety of materials. Additionally, it has been recognized that powder mixtures can be mechanically activated to induce chemical reactions, i.e., mechanochemical reactions at room temperature or at least at much lower temperatures than normally required to produce pure alloys and a variety of commercially useful materials. Efforts were also under way to understand the process fundamentals of MA through modeling studies.

Two different terms are commonly used to denote the processing of powder particles in high-energy ball mills. Mechanical Alloying describes the process when mixtures of powders (of different metals or alloys/compounds) are milled together. Material transfer is involved in this process to obtain a homogeneous alloy. On the other hand, mechanical milling of stoichiometric composition powders, such as intermetallics, or prealloyed powders, where material transfer is not required for homogenization, has been termed High Energy Ball Milling (HEBM).

The advantage of HEBM over MA is that since the powders are already alloyed and only a reduction in particle size and/or other transformations need to be induced mechanically, the time required for processing is short. For example, HEBM requires half the time required for MA to achieve the same effect. Additionally, HEBM of powders reduces oxidation of the constituent powders, related to the shortened time of processing. The technique of MA to synthesize novel alloy phases and to produce hydrogen storage nanomaterials has attracted the attention of a large number of researchers.

Mechanical alloying is a complex process. Some of the important parameters that have an effect on the final phase and/or microstructure of the product are: type of mill, material of the milling container, milling speed, milling time, size distribution of the grinding medium, ball-to-powder weight ratio, milling atmosphere and milling temperature.

In this part of the chapter, we concentrate generally on nanoprocessing conducted entirely in a solid state. This is a class of methods based on processes conducted in high-energy ball mills. Such methods are based on high-energy grinding and milling of materials. They provide a top-down approach toward manufacturing nanomaterials.

Figure 4.5 shows the SPEX Model 8,000 high-energy mill manufactured by the CertiPrep Company (Metuchen, NJ, USA). The vials (Fig. 4.6) are made of hardened ferritic steel, but vials made of zirconia, alumina, agate, and hard-metal tungsten carbide are also used to greatly limit contamination by grinding media, mainly iron [44]. The milling that yields nanometric structure takes place about 10 times faster in a SPEX ball mill than in a planetary ball mill. The milling time needed to produce fine powders and nanostructure depends obviously on the nature of material, but also on other parameters, such as the weight of the milling balls and the ratio of the weight of balls to the weight of powder (BPR). The ball-to-powder weight ratio can vary from 1 to 100. The time needed for milling to the same fines of a powder usually decreases with the increase in the BPR. Agglomeration and cold welding of metal particulates to the reactor wall and the milling balls or to form balls made of the milled powder itself are common phenomena that occur during milling of pure ductile metals in vials filled with inert gas, such as argon.

Figure 4.5 SPEX ball mill

Figure 4.6 Steal vial for SPEX mill.

4.3.1.2 Nanoprocessing

Five major processing methods can be employed during milling of materials in high-energy ball mills: (i) high-energy ball milling (HEBM), (ii) mechanical alloying (MA), (iii) mechanochemical activation synthesis (MCAS), (iv) mechanochemical synthesis (MCS), and (v) mechanical amorphization (MAM) [15].

High-energy ball milling is nanotechnology top-down approach for the synthesis of nanoparticles. In the HEBM process brute force is applied to material, whether it is a metal, a pre-alloyed intermetallic, or a solid chemical (stoichiometric) compound. Short milling times can break thin chemically passive surface coatings (e.g., surface oxides) and expose fresh, clean chemically active metallic surface. Such milling can also introduce defects into solid compounds. It results in an increased chemical activity of milled media toward both gasses and chemical reactions in solutions and electrolytes. This process is often termed mechanochemical activation synthesis.

A process wherein mixture of elemental metal powders, or powders of metal and nonmetal are milled long enough to trigger alloying of elemental powders in a solid-state process is termed mechanical alloying. Since substantial rearrangement of chemical species must take place during alloying the latter requires long times of milling, usually more than twice the time needed for preparation of nanopowders by mechanical milling. Even longer milling results in what we can term mechanical amorphization of a crystalline solid.

Reactive mechanical milling can be realized by reacting two chemical compounds A and B in a solid-state synthesis that yields a distinct chemical compound C. This is termed mechanical synthesis or mechanosynthesis. Reactive milling can also be conducted by milling metal powders in ball mills filled with a reactive gas.

4.3.1.3 Mechanical alloying

Mechanical alloying, an important mechanical process, was developed in the 1970s at the International Nickel Co as a technique for dispersing nanosized inclusions into nickel-based alloys (Fig. 4.7) [1, 3]. During the last years, the MA process has been successfully used to prepare a variety of alloy powders including powders exhibiting supersaturated solid solutions, quasicrystals, amorphous phases, and nano-intermetallic compounds [4, 5]. The MA technique has proved to be a novel and promising method for alloy formation [2, 6–16].

Mechanical alloying is conducted through milling of two elemental metal powders. Longer milling, order of hours, leads to atomic-level mixing of the metals and produces an alloy consisting of these metals. Until the advent of this new, nonequilibrium and low-temperature solid-state processing method, metal alloys could only be manufactured by melt casting and metal foundry practices.

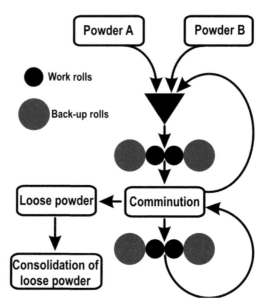

Figure 4.7 Mechanical alloying process.

The raw materials used for MA are available commercially high-purity powders that have sizes in the range of 1–100 μm. During the mechanical alloying process, the powder particles are periodically trapped between colliding balls and are plastically deformed. Such a feature occurs by the generation of a wide number of dislocations as well as other lattice defects. Furthermore, the ball collisions cause fracturing and cold welding of the elementary particles, forming clean interfaces at the atomic scale. Further milling lead to an increase of the interface number and the sizes of the elementary component area decrease from millimeter to submicrometer lengths. Concurrently to this decrease of the elementary distribution, some nanocrystalline intermediate phases are produced inside the particles or at its surfaces. As the milling duration develops, the content fraction of such intermediate compounds increases leading to a final product whose properties are the function of the milling conditions.

The sequence of concurrent mechanical and chemical events can be written as follows [15]: mechanical stressing → severe plastic deformation → formation of a submicron lamellar microstructure → cold interdiffusion of metal atoms between lamellae or nanograins (cold welding) → fracture → formation of nanostructure → extended solid solubility → mechanical alloying with formation of thermodynamically stable and/or metastable phases → amorphization.

It was shown that mechanical alloying has produced amorphous phases in metals. But differentiation between a "truly" amorphous, extremely fine grained, or a material in which very small crystals are embedded in an amorphous matrix in so produced materials has not been easy on the basis of diffraction basis. Only the supplementary investigations by neutron diffraction can unambiguously confirmed that the phases produced by MA are truly amorphous. The milled powder is finally heat treated to obtain the desired microstructure and properties. Annealing leads to grain growth and release of microstrain.

Nanocrystalline materials exhibit quite different properties from both crystalline and amorphous materials, due to structure, in which extremely fine grains are separated by what some investigators have characterized as "glass-like" disordered grain boundaries. The mechanism of amorphous phase formation by MA is due to a chemical solid state reaction, which is believed to be caused by the formation of a multilayer structure during milling [5, 15]. When a mixture of elemental powders is milled, the formation of the amorphous phase is mainly due to an ultimate interdiffusion of atoms that occurs at fresh surfaces and interfaces created by mechanical milling. This interdiffusion is promoted by defects and chemical disorder in the crystalline structure.

In Fig. 4.8, the correlation between the different parameters of MA process is presented. The basic process of mechanical alloying is illustrated in Fig. 4.9. Microcrystalline powders are placed together with a number of hardened steel or tungsten carbide (WC) coated balls in a sealed container which is shaken. The most effective ratio for the ball-to-powder masses is 10.

Furthermore, for all nanocrystalline materials prepared by a variety of different synthesis routes, surface and interface contamination is a major concern. In particular, during mechanical attrition, contamination by the milling tools (iron) and atmosphere (trace elements of O_2 and other rare gases) can be a problem (see Fig. 4.10) [34–37]. By minimizing the milling time and using the purest, most ductile metal powders available, a thin coating of the milling tools by the respective powder material can be obtained which reduces Fe contamination tremendously. Atmospheric contamination can be minimized or eliminated by sealing the vial with a flexible "O" ring after the powder has been loaded in an inert gas glove box.

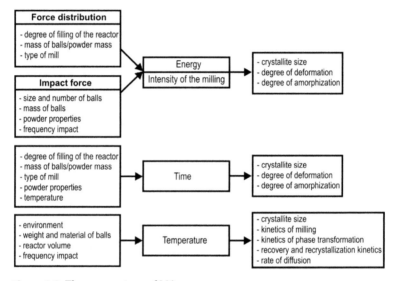

Figure 4.8 The parameters of MA process.

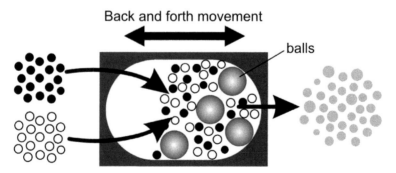

Figure 4.9 Schematic cross-sectional representation of MA process for synthesizing nanometer-sized powders (SPEX 8000 mixer mill).

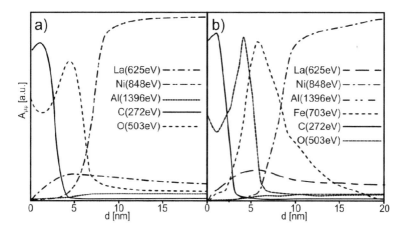

Figure 4.10 AES spectrum of microcrystalline (a) and nanocrystalline (b) LaNi$_{4.2}$Al$_{0.8}$ alloy vs. sputtering time, as converted to depth. The sample surface is located on the left-hand side.

4.4 XPS and EAS Studies

The cleanliness of the surface of nanocrystalline alloys prepared by mechanical alloying method was studied by X-ray photoelectron spectroscopy (XPS) and Auger electron spectroscopy (AES) [35]. In Fig. 4.10, we show the element specific Auger intensities of the nanocrystalline LaNi$_{4.2}$Al$_{0.8}$ sample as a function of the sputtering time, converted to depth. Data for microcrystalline LaNi$_{4.2}$Al$_{0.8}$ sample are included, too.

As can be seen there is relatively high concentration of carbon and oxygen immediately on the surface, which could be due to carbonates or adsorbed atmospheric CO_2. The carbon concentration strongly decreases toward the interior of the sample. At the metal interface itself, only oxygen is present, making it very likely that only an oxide layer is formed, and no other compounds, the later growing apparently with a smaller probability. We have found that at the oxide–metal interface mainly lanthanum and nickel atoms are present. Taking into account that the escape-depth of the Auger electron from nickel and aluminum atoms is about 2 nm, the concentration of these elements on the metallic surface is significantly lower compared to the average bulk composition. Therefore, the lanthanum atoms, which segregate to the surface form a La-based oxide layer under atmospheric conditions. The oxidation process is depth limited such that an oxide-covering layer with a well-defined thickness is formed by which the lower lying metal is prevented from further oxidation. In this way one can obtain a self-stabilized oxide–metal structure. Very similar behavior was observed for the microcrystalline Co thin films oxidized under atmospheric conditions. From the peak-to-peak amplitude, we have calculated a maximum atomic concentration of oxygen inside the sample as 2 at%. The concentration of carbon impurities inside the sample was below 0.5 at%.

Figure 4.10b shows the element specific Auger intensities of the nanocrystalline $LaNi_{4.2}Al_{0.8}$ sample as a function of the sputtering time, converted to depth. Similar to the results obtained for the microcrystalline sample, there is relatively high concentration of carbon and oxygen immediately on the surface. The carbon concentration strongly decreases toward the interior of the sample. We have found that at the oxide–metal interface only iron impurities and lanthanum atoms are present. As the escape-depth of the Auger electrons from nickel and aluminum atoms is about 2 nm, we conclude that these elements are practically not present on the metallic surface. In other words, lanthanum atoms and iron impurities strongly segregate to the surface and form an oxide layer under atmospheric conditions. The lower lying Ni atoms form a metallic subsurface layer and are responsible for the observed high hydrogenation rate in accordance with earlier findings. The above segregation process is stronger compared to that observed for the microcrystalline sample. The presence of the significant amount of iron atoms in the surface layer of the nanocrystalline $LaNi_{4.2}Al_{0.8}$ alloy could be explained by Fe impurities trapped in the MA powders from erosion of the milling media. The amount of the Fe impurities considerably decreases in the subsurface layer of the sample. From the peak-to-peak amplitude, we have estimated a maximum atomic concentration of oxygen as ~2 at% in the interior of the sample. Similarly to the results reported for the polycrystalline $LaNi_{4.2}Al_{0.8}$ alloy, the concentration of carbon impurities inside the sample was below 0.5 at%.

Concluding, the contamination with Fe based wear debris in mechanically synthesized nanomaterials can generally be reduced to less than 1–2% and oxygen contamination to less than 300 ppm. However, contamination through the milling atmosphere can have a positive impact on the milling conditions if one wants to prepare metal or ceramic nanocomposites with one of the metallic elements being chemically highly reactive with the gas (or fluid) environment. On the other side, main advantage of the top-down approach can be high production rates of nanopowders.

References

1. Benjamin, J. S. (1976). Mechanical alloying. *Sci. Am.* **234**: 40–57.
2. Zaluski L., Zaluska A., and Ström-Olsen J. O. (1995). Hydrogen absorption in nanocrystalline Mg_2Ni formed by mechanical alloying. *J. Alloys Comp.* **217**: 245–249.
3. Benjamin, J. S. (1997). Mechanical alloying process (United States Patent 5688303).
4. Suryanarayana, C., and Koch, C. C. (1999). Nanostructured materials, in *Non-Equilibrium Processing of Materials*, Suryanarayana, C., ed. (Elsevier Science Pub., Oxford, UK), pp. 313–346.
5. Suryanarayna, C. (2001). Mechanical alloying. *Prog. Mater. Sci.* **46**: 1–184.
6. Zaluska, A., Zaluski, L., Ström-Olsen, J. O. (2001). Structure, catalysis and atomic reactions on the nano-scale: A systematic approach to metal hydrides for hydrogen storage. *Appl. Phys. A* **72**: 157–165.
7. Jurczyk, M., Nowak, M., Jankowska, E., and Gasiorowski, A. (2001). Nanocrystalline alloy hydride electrodes formed by mechanical alloying. *Mater. Sci. Eng.* **4**: 416–410.
8. Jurczyk, M., Jankowska, E., Nowak, M., and Jakubowicz, J. (2002). Nanocrystalline titanium type metal hydrides prepared by mechanical alloying. *J. Alloys Comp.* **336**: 265–269.
9. Jurczyk, M., Jankowska, E., Nowak, M., and Wieczorek, I. (2003). Electrode characteristics of nanocrystalline TiFe-type alloys. *J. Alloys Comp.* **354**: L1–L4.
10. Jurczyk, M. (2004). The progress of nanocrystalline hydride electrode materials. *Bull. Pol. Acad. Sci. Tech. Sci.* **52**: 67–77.
11. Jurczyk, M., Smardz, L., Makowiecka, M., Jankowska, E., and Smardz, K. (2004). The synthesis and properties of nanocrystalline electrode materials by mechanical alloying. *J. Phys. Chem. Sol.* **65**: 545–548.

12. Jurczyk, M., Smardz, L., and Szajek, A. (2004). Nanocrystalline materials for NiMH batteries. *Mater. Sci. Eng. B* **108**: 67–75.
13. Gasiorowski, A., Iwasieczko, W., Skoryna, D., Drulis, H., and Jurczyk, M. (2004). Hydriding properties of nanocrystalline $Mg_{2-x}M_xNi$ alloys synthesized by mechanical alloying (M = Mn, Al). *J. Alloys Comp.* **364**: 283–288.
14. Jurczyk, M., and Nowak, M. (2008). Nanomaterials for hydrogen storage synthesized by mechanical alloying, in: *Nanostructured Materials in Electrochemistry*, Ali, E., ed. (Wiley), Chapter 9.
15. Varin, R. A., Czujko, T., and Wronski, Z. S. (2009). *Nanomaterials for Solid State Hydrogen Storage*, Springer Science+Business Media, LLC.
16. Li, X. D., Elkedim, O., Nowak, M., Jurczyk, M., and Chassagnon, R. (2013). Structural characterization and electrochemical hydrogen storage properties of $Ti_{2-x}Zr_xNi$ (x = 0, 0.1, 0.2) alloys prepared by mechanical alloying. *Int. J. Hydrogen Energy* **38**: 12126–12132.
17. Balcerzak, M., Jakubowicz, J., Kachlicki, T., and Jurczyk, M. (2015). Effect of multi-walled carbon nanotubes and palladium addition on the microstructural and electrochemical properties of the nanocrystalline Ti_2Ni alloy. *Int. J. Hydrogen Energy* **40**: 3288–3299.
18. Balcerzak, M., Jakubowicz, J., Kachlicki, T., and Jurczyk, M. (2015). Hydrogenation properties of nanostructured Ti_2Ni-based alloys and nanocomposites. *J. Power Sources* **280**: 435–445.
19. Anani, A., Visintin, A., Petrov, K., and Srinivasan, S. (1994). Alloys for hydrogen storage in nickel/hydrogen and nickel/metal hydride batteries. *J. Power Sources* **47**: 261–275.
20. Bradhurst, D. H. (1983). Metal hydrides for energy storage. *Metals Forum* **6**: 139–148.
21. Principi, G., Agresti, F., Maddalena, A., and Lo Russo, S. (2009). The problem of solid state hydrogen storage. *Energy* **34**: 2087–2091.
22. Jain, I. P. (2009). Hydrogen the fuel for 21st century. *Int. J. Hydrogen Energy* **34**: 7368–7378.
23. Zhao, X. G., and Ma, L. Q. (2009). Recent progress in hydrogen storage alloys for nickel/metal hydride secondary batteries—review. *Int. J. Hydrogen Energy* **34**: 4788–4796.
24. Jain, I. P., Jain, P., and Jain, A. (2010). Novel hydrogen storage materials: A review of lightweight complex hydrides. *J. Alloys Comp.* **503**: 303–339.
25. Young, K. H., and Nei, J. (2013). The current status of hydrogen storage alloy development for electrochemical applications. *Materials* **6**: 4574–4608.

26. Dowling, A. P. (2004). Development of nanotechnologies, *Mater. Today* **7**, pp. 30–35.
27. Scheu, M., Veefkind, V., Verbandt, Y., Galan, M. G., Absalom, R., and Förster, W. (2006). Mapping nanotechnology patents: The EPO approach, *World Patent Inf.*, **28**, 204–211.
28. Zweck, A., Bachmann, G., Luther, W., and Ploetz, C. (2008). Nanotechnology in Germany: From forecasting to technological assessment to sustainability studies, *J. Cleaner Prod.* **16**, 977–987.
29. Kleparis, J., Wojcik, G., Czerwinski, A., Skowronski, J., Kopczyk, M., Beltowska-Brzezinska, M. (2001). Electrochemical behavior of metal hydrides, *J. Solid State Electrochem.* **5**, 229–249.
30. Kopczyk, M., Wojcik, G., Młynarek, G., Sierczynska, A., and Beltowska-Brzezinska, M. (1996). Electrochemical absorption-desorption of hydrogen on multicomponent Zr-Ti-V-Ni-Cr-Fe alloys in alkaline solution, *J. Appl. Electrochem.* **26**, pp. 639–645.
31. Sakintuna, B., Lamari-Darkrim, F., Hirscher, M. (2007). Metal hydride materials for solid hydrogen storage: A review, *Int. J. Hydrogen Energy* **32**, 1121–1140.
32. Kalinichenka, S., Röntzsch, R., and Kieback, B. (2009). Structural and hydrogen storage properties of melt-spun Mg–Ni–Y alloys. *Int. J. Hydrogen Energy* **34**, 7749–7755.
33. Zadorozhnyi, V. Yu., Skakov, Yu, A., and Milovzorov, G. S. (2008). Appearance of metastable states in Fe–Ti and Ni–Ti systems in the process of mechanochemical synthesis, *Metal Sci. Heat Treat.* **50**, 404–410.
34. Jurczyk, M., Smardz, L., and Szajek, A. (2004). Nanocrystalline materials for NiMH batteries, *Mater. Sci. Eng. B* **108**, 67–75.
35. Smardz, L., Jurczyk, M., Smardz, K., Nowak, M., Makowiecka, M., and Okonska, I. (2008). Electronic structure of nanocrystalline and polycrystalline hydrogen storage materials, *Renew. Energy* **33**, 201–210.
36. Smardz, K., Smardz, L., Okonska, I., Nowak, M., and Jurczyk, M. (2008). XPS valence band and segregation effect in nanocrystalline Mg_2Ni-type materials. *Int. J. Hydrogen Energy* **33**: 387–392.
37. Smardz, L., Smardz, K., Nowak, M., and Jurczyk, M. (2001). Structure and electronic properties of $La(Ni,Al)_5$ alloys. *Cryst. Res. Tech.* **36**: 1385–1392.

Chapter 5

X-Ray Diffraction

Maciej Tulinski

*Institute of Materials Science and Engineering,
Poznan University of Technology, Poznan, Poland*

maciej.tulinski@put.poznan.pl

5.1 Introduction

This chapter deals with a broad subject, including crystallography, production, detection, characteristics, and practical use of X-rays. The described methods relate primarily to the study of polycrystalline materials, and studies of monocrystals have been deliberately excluded.

First, a brief overview of crystallography is presented: how atoms are arranged in space of crystals and how it is mathematically described. Also, it is explained how the orientation of planes in crystals can be scientifically represented. Subsequently, the role and classification of defects in crystals are presented.

Further, the chapter deals with X rays, how they are produced, and how they can be detected.

The last part of the chapter is devoted to the practical use of diffraction, including measurement of basic parameters, such as crystallite size.

Handbook of Nanomaterials for Hydrogen Storage
Edited by Mieczyslaw Jurczyk
Copyright © 2018 Pan Stanford Publishing Pte. Ltd.
ISBN 978-981-4745-66-6 (Hardcover), 978-1-315-36444-5 (eBook)
www.panstanford.com

Only the basic aspects of above-mentioned topics are covered here, and readers are encouraged to deepen their knowledge in X-ray diffraction studying books and scientific articles included in the bibliography at the end of this chapter.

5.2 Geometry of Crystals

A solid composed of atoms, ions or molecules arranged in a periodic pattern is called a crystal [1]. Due to the requirement of periodicity, crystals are different from gases and liquids. Depending on the crystallization conditions, they can be single-crystals or consist of many contiguous crystals—the latter are called polycrystals. These solids have many interesting electrical, magnetic, optical, and mechanical properties, which differ from other solids—called amorphous (Fig. 5.1) [2].

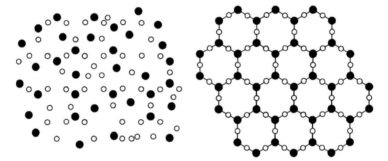

Figure 5.1 Amorphous structure (left) and crystalline solid (right).

5.2.1 Lattices and Crystal Systems

The structure of all the crystals can be described by the lattice (Fig. 5.2). It is a mathematical abstraction, in which each point (node) is associated with a group of atoms called the base.

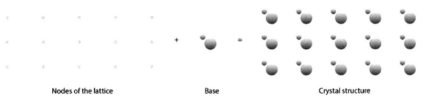

Figure 5.2 Formation of the crystal structure.

Crystal lattices can be mapped into themselves by the lattice translations and symmetry operations, i.e., rotation about an axis that passes through a lattice point. There are an unlimited number of possible lattices, because there is no restriction on the length of the lattice translation vectors or on the angle between them. The point symmetry groups in three dimensions require 14 different lattice types (see Table 5.1) [3].

Table 5.1 Different lattice types in three dimensions

System	Number of lattices	Cell axial lengths	Angles
Triclinic	1	$a \neq b \neq c$	$A \neq \beta \neq \gamma$
Monoclinic	2	$A \neq b \neq c$	$A = \gamma = 90° \neq \beta$
Orthorhombic	4	$a \neq b \neq c$	$a = \beta = \gamma = 90°$
Tetragonal	2	$a = b \neq c$	$A = \beta = \gamma = 90°$
Cubic	3	$a = b = c$	$a = \beta = \gamma = 90°$
Trigonal	1	$a = b = c$	$A = \beta = \gamma < 120° \neq 90°$
Hexagonal	1	$a = b \neq c$	$a = \beta = 90° \gamma = 120°$

Seven different point lattices can be obtained simply by putting points at the corners of the unit cells of the seven crystal systems. However, there are other arrangements of points that fulfill the requirement that each lattice point have identical surroundings. That leads to fourteen lattice types (called Bravais lattices). They are illustrated in Fig. 5.3.

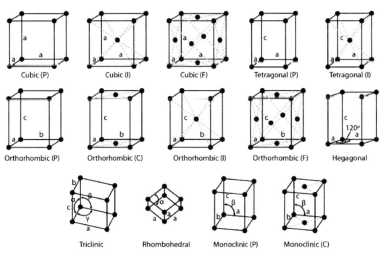

Figure 5.3 Bravais lattices in three dimensions.

For example, there are three lattices in the cubic system:
- simple cubic lattice (P, sc)
- body-centered cubic lattice (I, bcc)
- face centered cubic lattice (F, fcc)

Primitive cells (symbol P and R) have only one lattice point per cell, while non-primitive cells have more than one.

5.2.2 Designation of Planes

The orientation of planes in a lattice can be represented symbolically, using system proposed by crystallographer Miller. Miller indices can be determined by finding the intercepts on the axes in terms of the lattice constants a, b, c, taking the reciprocals of these numbers and reducing to three integers having the same ratio, usually the smallest integers. As a result, the index of the plane is written in parentheses (hkl). If a plane is parallel to a given axis, its fractional intercept on that axis is taken as infinity and the corresponding Miller index is zero. If a plane cuts a negative axis, the corresponding index is negative and is written with a bar over it [4].

An example of determining the Miller indices of the plane is shown in Fig. 5.4.

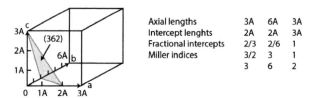

Figure 5.4   Determining the Miller indices of the plane.

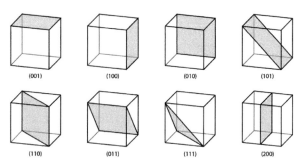

Figure 5.5   Miller indices of lattice planes.

A few examples of designation of Miller indices are presented in Fig. 5.5.

The spacing d between adjacent (hkl) lattice planes can be calculated using equations from Table 5.2.

Table 5.2 Basic equations

System	Interplanar distance d_{hkl}
Cubic	$\dfrac{1}{d^2} = \dfrac{h^2 + k^2 + l^2}{a^2}$
Tetragonal	$\dfrac{1}{d^2} = \dfrac{h^2 + k^2}{a^2} + \dfrac{l^2}{c^2}$
Rhombohedral	$\dfrac{1}{d^2} = \dfrac{(h^2 + k^2 + l^2)\sin^2\alpha + 2(hk + kl + hl)(\cos^2\alpha - \cos\alpha)}{a^2(1 - 3\cos^2\alpha + 2\cos^3\alpha)}$
Hexagonal	$\dfrac{1}{d^2} = \dfrac{4}{3}\dfrac{h^2 + hk + k^2}{a^2} + \dfrac{l^2}{c^2}$
Orthorombic	$\dfrac{1}{d^2} = \dfrac{h^2}{a^2} + \dfrac{k^2}{b^2} + \dfrac{l^2}{c^2}$
Monoclinic	$\dfrac{1}{d^2} = \dfrac{h^2}{a^2\sin^2\beta} + \dfrac{k^2}{b^2} + \dfrac{l^2}{c^2\sin^2\beta} - \dfrac{2hl\cos\beta}{ac\sin^2\beta}$
Triclinic	$\dfrac{1}{d^2} = \dfrac{\left(\dfrac{h^2\sin^2\alpha}{a^2} + \dfrac{k^2\sin^2\beta}{b^2} + \dfrac{l^2\sin^2\gamma}{c^2}\right)}{(1 - \cos^2\alpha - \cos^2\beta + \cos^2\gamma + 2\cos\alpha\cos\beta\cos\gamma)}$

5.2.3 Defects in Crystals

There is no perfect crystal in the nature. In real materials, atom arrangements do not follow perfect crystalline patterns. There are a number of imperfections and it is useful to classify them by their dimension.

The zero-dimensional defects affect isolated sites in the crystal structure and are hence called point defects. An example is a solute or impurity atom, which alters the crystal pattern at a single point. An intrinsic defect is formed when an atom is missing from a position that ought to be filled in the crystal, creating a vacancy, or when an atom occupies an interstitial site where no atom would ordinarily appear, causing an interstitialcy. The two types of intrinsic point defects are shown in Fig. 5.6.

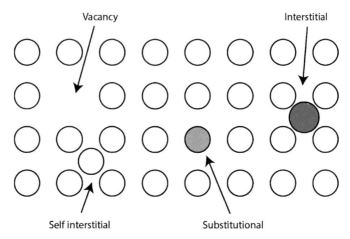

Figure 5.6 Point defects.

The one-dimensional defects are called dislocations. Basically they are lines along which the crystal pattern is broken. Edge and screw dislocations are presented in Fig. 5.7. They play a role in the microstructure controlling the yield strength and subsequent plastic deformation of crystalline solids at ordinary temperatures. Dislocations also participate in the growth of crystals and in the structures of interfaces between crystals [5].

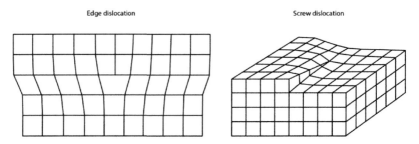

Figure 5.7 Linear defects.

The two-dimensional defects are surfaces, such as the external surface and the grain boundaries (Fig. 5.8) along which distinct crystallites are joined together [6].

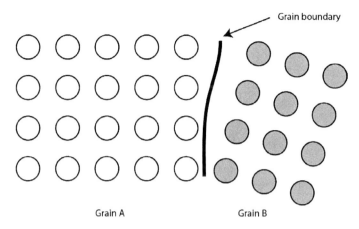

Figure 5.8 Grain boundary.

The three-dimensional defects change the crystal pattern over a finite volume. They include precipitates, which are small volumes of different crystal structure, and also include large voids or inclusions of second-phase particles.

5.3 Properties of X-Rays

X-ray radiation is an electromagnetic radiation of a wavelength between 1 pm to 10^4 pm. X-ray wavelength most commonly is expressed in Ångströms (1 Å = 10^{-10} m). Thus, the above-mentioned range is about 10^{-2}–10^2 Å. The nature of this radiation is the same as, for example, Gamma radiation; the difference is only in the wavelength and the method of formation. X rays were discovered in 1895 by physicist Wilhelm Röntgen [7].

5.3.1 Electromagnetic Radiation

Electromagnetic radiation is a form of radiant energy released by certain electromagnetic processes. Classically, electromagnetic radiation consists of electromagnetic waves, which are synchronized oscillations of electric and magnetic fields that propagate at the speed of light. Electromagnetic waves can be characterized by either the frequency (ν) or wavelength (λ) of their oscillations to form the electromagnetic spectrum (Fig. 5.9), which includes, in order of increasing frequency and decreasing wavelength:

radio waves, microwaves, infrared radiation, visible light, ultraviolet radiation, X-rays, and gamma rays [8].

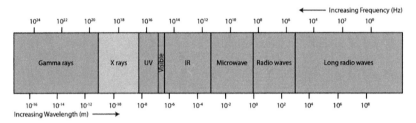

Figure 5.9 The electromagnetic spectrum.

5.3.2 The Continuous Spectrum

Total X-ray intensity I is given by

$$I_{\text{cont.spec.}} = aiZU^m, \tag{5.1}$$

where a is the proportionality constant, i the tube current, Z the atomic number, U the applied voltage, and m the constant with a value of about 2.

Thus, total X-ray intensity increases with

- increasing the tube current;
- increasing the atomic number of tube's target;
- increasing applied voltage.

This relationship can be seen in Fig. 5.10.

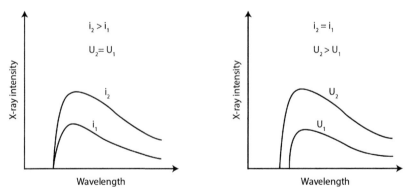

Figure 5.10 The influence of the applied voltage and the tube current on the intensity of the emitted radiation.

The produced radiation is called polychromatic, continuous, or white radiation. The continuous spectrum results from the rapid deceleration of the electrons hitting the target—that is why it is also called *Bremsstrahlung* (German for "braking radiation") [9].

5.3.3 The Characteristic Spectrum

Sharp intensity maxima that appear on the continuous spectrum at certain wavelengths are called characteristic lines. They are characteristic of the target metal and they appear when the voltage on an X-ray tube is raised above a certain critical value. Characteristic lines fall into several sets, in the order of increasing wavelength, and are referred to as *K*, *L*, *M*, and so on. All the lines together are forming the characteristic spectrum of the metal used as the target (Fig. 5.11).

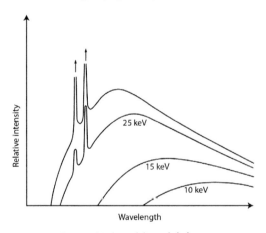

Figure 5.11 X-ray spectrum of tube with molybdenum target.

The intensity of the characteristic line depends on the tube current *i* and the amount by which the applied voltage *U* exceeds the critical excitation voltage for that line. For example, for a K line, the intensity is given by

$$I_K = bi(U - U_K)^n, \qquad (5.2)$$

where *b* is the proportionality constant, U_K the K excitation voltage, and *n* the constant of 1.5 for K series.

Characteristic lines are very narrow (most of them less than 0.001 Å) and very intense. In practice, only the K lines are useful in X-ray diffraction. They consist of four most intense lines: α_1, α_2, β_1, and β_2. Wavelengths of K lines for some X-ray tubes are given in Table 5.3.

Table 5.3 Wavelengths of K lines for common X-ray tubes

Element	Atomic number Z	α_2	α_1	β_1	β_2
Cr	24	2.293	2.289	2.084	—
Mn	25	2.105	2.101	1.910	—
Fe	26	1.939	1.935	1.756	—
Co	27	1.792	1.788	1.620	—
Ni	28	1.661	1.657	1.500	1.488
Cu	29	1.544	1.540	1.392	1.381
Mo	42	0.713	0.709	0.632	0.620
Ag	47	0.563	0.559	0.497	0.487
W	74	0.213	0.208	0.184	0.179

Relationship (5.2) is valid for voltages $U < U_K$. In case of voltage $5U_K < U < 11U_K$, there is slight increase in intensity of characteristic line, and for $U > 11U_K$ there is decrease of that value. Therefore, the voltage applied to the X-ray tube is usually four or five times higher than U_K.

5.3.4 Production of X-Rays

There are many sources of X-rays: X-ray tubes, radioactive isotopes, or accelerators. The most commonly used are X-ray tubes. They contain two electrodes, an anode (target), and a cathode maintained at a high negative potential. Electrons are provided by the ionization of a small quantity of gas (gas tubes, now obsolete) or from hot filament (filament tubes) [10].

Filament tubes consist of an evacuated glass container, an anode, and a cathode (Fig. 5.12). The cathode is in form of a tungsten filament and the anode is (in most cases) a water-cooled block of copper, containing the desired target metal.

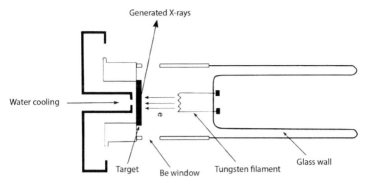

Figure 5.12 Cross section of X-ray tube.

Electrons emitted by the cathode are accelerated by the electric field between the electrodes. As a result of the acceleration, the electrons obtain kinetic energy. They are decelerated at the anode, so that they lose energy and produce a number of independent phenomena, including

- the creation of bremsstrahlung, characterized by a continuous spectrum.
- the electrons with high kinetic energy excite the atoms of the material of the anode through the knocking of electrons out of the shell. Upon return to normal state atom radiate excess energy in the form of X-ray quanta (characteristic radiation with linear spectrum);
- the electrons with a high kinetic energy change the position of atoms of the anode material to not-equilibrium, increasing the amplitude of their vibrations. The result is an increase in the temperature of the anode.

Thus, the X-ray tube produces radiation with continuous spectrum and characteristic radiation with linear spectrum, wherein both the spectrums are additively superimposed on each other.

5.3.5 Detection of X-Rays

The X-ray diffractometers use different detectors: Geiger–Müller, proportional, scintillation, and semiconductor.

Geiger–Müller and proportional detector are called gas detectors (or counters) and they are based on ionization phenomenon. They consist of a cylindrical metal shell (the cathode) filled with gas and containing a metal wire (the anode) (see Fig. 5.13). The X-rays that enter the cylinder are absorbed by the gas and photoelectrons and Compton recoil electrons are ejected. As a result, ionization of the gas occurs and electrons are produced. These electrons move under the influence of the electric field toward the anode. Also produced positive gas ions move toward the cathode. All these electrons and ions will be collected on the electrodes and there will be a small current, which is a measure of the X-ray intensity.

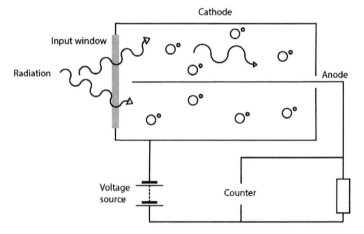

Figure 5.13 Geiger counter.

At a potential difference of about 200 V, this device is called an ionization chamber. If the voltage is raised to approximately 1000 V, the same instrument acts as a proportional counter. In this case, electrons produced by the primary ionization acquire enough energy to knock electrons out of other gas atoms, which cause further ionization and in turn an easily detectable current in the circuit. The size of the current is proportional to the energy of the X-ray quantum absorbed (hence the name of the counter).

Voltage on a proportional counter can be increased, and above 1500 V it acts as a Geiger–Müller counter. In this case, some atoms are ionized by the applied voltage, but also other atoms are raised to excited states and emits ultraviolet radiation, which is also knocking electrons out. As a result, avalanches of electrons are produced and reaches the anode.

Another type of detector utilizes substances that fluoresce visible light when exposed to the X-rays (Fig. 5.14). The amount of emitted light is proportional to the X-ray intensity. Every absorbed X-ray quantum produces a flash of light (scintillation) in the crystal (NaI is mostly used). Then the light passes to the photocathode and ejects a number of electrons from it. The resulting electrons are drawn to the dynodes, electrons knocks another electrons out from the metal surface and finally measureable pulses, proportional to the energy of the absorbed X-ray quanta, are produced.

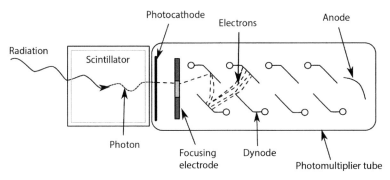

Figure 5.14 Schematic of scintillation detector.

Semiconductor detectors produce pulses proportional to the absorbed X-ray energy. In these detectors, fundamental information carriers are electron–hole pairs, which are produced along the path taken by the charged particle through the detector. By collecting electron–hole pairs, the detection signal is formed.

Of the available semiconductor materials, silicon is mainly used for charged particle detectors and soft X-ray detectors while germanium is widely used for gamma-ray spectroscopy.

Thermal excitation in semiconductor materials is strongly dependent on temperature. A pure semiconductor is an insulator at absolute zero and as the temperature increases electrons can be thermally excited from the valence band to the conduction band.

Ideal radiation detectors should have no charge in the absence of radiation (and lots of charge in the presence of an ionizing radiation event). This is one of the reasons why some semiconductor detectors are cooled with liquid nitrogen. Cooling reduces the number of electron–hole pairs in the crystal.

5.4 Diffraction

Diffraction is essentially a scattering phenomenon, in which a large number of atoms cooperate. The atoms are arranged periodically on a lattice and the rays scattered by them have definite phase relations between them.

5.4.1 Bragg's Law

Bragg diffraction was first proposed by William Lawrence Bragg and William Henry Bragg in 1913 in response to their discovery that crystalline solids produced surprising patterns of reflected X-rays. They found that in these crystals, for certain specific wavelengths and incident angles, intense peaks of reflected radiation (known as Bragg peaks) were produced. W. L. Bragg explained this result by modeling the crystal as a set of discrete parallel planes separated by a constant parameter d. It was proposed that the incident X-ray radiation would produce a Bragg peak if their reflections off the various planes interfered constructively. This concept of diffraction applies equally to neutron diffraction and electron diffraction processes [11].

Bragg diffraction occurs when electromagnetic radiation or subatomic particle waves with wavelength comparable to atomic spacings d are incident upon a crystalline sample, scattered by the atoms in the system and undergo constructive interference.

One has to consider conditions necessary to make the phases of the beams coincide when the incident angle equals reflecting angle (Fig. 5.15). The rays of the incident beam are always in phase and parallel up to the point at which the top beam strikes the top layer at atom A. The second beam continues to the next, parallel layer where it is scattered by atom B. The second beam must travel the extra distance CB + BD if the two beams are to continue traveling adjacent and parallel. This extra distance must be an integral (n) multiple of the wavelength (λ) for the phases of the two beams to be the same:

$$n\lambda = CB + BD \qquad (5.3)$$

Recognizing d as the hypotenuse of the triangle ABC, we can use trigonometry to relate d and θ to the distance (CB + BD). The distance CB is opposite to θ, thus

$$CB = d\sin\theta. \tag{5.4}$$

Because CB = BD, Eq. (5.3) becomes

$$n\lambda = 2CB. \tag{5.5}$$

Substituting Eq. (5.4) in Eq. (5.5) we obtain

$$n\lambda = 2d\cdot\sin\theta. \tag{5.6}$$

This relation is known as Bragg's law. It states the essential condition for occurring diffraction. The rays scattered by all the atoms in all planes are completely in phase and form a diffracted beam.

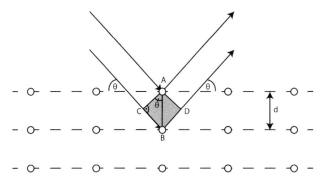

Figure 5.15 Diffraction of X-rays.

Bragg's law explains also the necessity of using X-rays. For most sets of crystal planes d is of the order of 3 Å or less. It means that λ cannot exceed about 6 Å. That is why a crystal could not diffract i.e., ultraviolet radiation (λ approx. 500 Å) or visible light (λ = 400–700 nm).

5.4.2 Laue's Equations

In 1912, Max von Laue analyzed the X-ray scattering at the nodes of the crystal lattice. He obtained mathematical relationships that described that phenomenon—named after him Laue's equations [12].

He assumed that the crystal has defect-free structure, the atoms are motionless (no thermal effects), X-rays are parallel and the crystal does not absorb them.

To simplify, consider a one-dimensional array of scatterers, spaced a apart (Fig. 5.16). The X-ray beam has a wavelength λ and makes an angle α_0 with the line of scatterers.

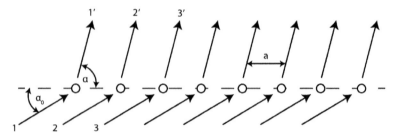

Figure 5.16 One-dimensional array of scatterers.

Let us consider rays no. 1 and 2 of the beam (Fig. 5.17).

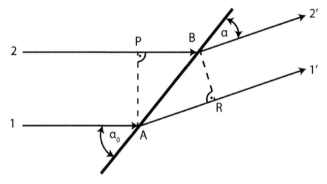

Figure 5.17 Scheme for deriving Laue's equation.

The straight line AP sets the front of the waves. The waves on straight lines AR and PB have different distances to go. The distance difference ΔS can be written as

$$\Delta S = AR - PB. \tag{5.7}$$

Diffracted rays 1' and 2' can interfere with each other if the distance difference is equal to an integral multiple (h) of wavelength (λ):

$$\Delta S = h\lambda \tag{5.8}$$

Equation (5.7) is equal to Eq. (5.8); thus

AR − PB = $h\lambda$. (5.9)

Taking into consideration:

AR = AB $\cos\alpha$ = $a \cos\alpha$ (5.10)

PB = AB $\cos\alpha_0$ = $a \cos\alpha_0$ (5.11)

the result is:

$a(\cos\alpha - \cos\alpha_0) = h\lambda$ (5.12)

Equation (5.12) is known as Laue's equation [12].

5.4.3 Crystallite Size

Given the three-dimensional nanomaterials, instead of the term "grain" we use the term "crystallite"—an area coherently scattering X-rays. A few crystallites joined together in porousless system is called aggregate, while the agglomerate provides a system interconnected in a porous way, consisting of the number of crystallites, as well as aggregates.

Figure 5.18 shows a two-dimensional structure of the nanocrystalline material, in which the crystallites have a different crystallographic orientation. Grain boundaries (open circles) have a lower density.

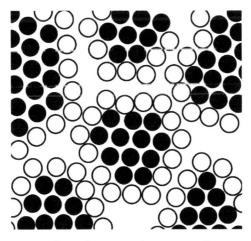

Figure 5.18 Structure of two-dimensional nanocrystalline material.

Grain is characterized by a specific chemical composition, crystalline structure and physicochemical and mechanical properties. In characterizing nanocrystalline materials, we pay attention to measurement techniques able to assess the size of the grains. Crystallite size can be determined with X-ray diffraction technique using the Scherrer and Hall method.

We use the following relation between the crystallite size and the width of Bragg reflexes: the smaller the grain size, the greater broadening of reflections. We determine the crystallite size based on the Scherrer equation [13]:

$$B = K\lambda/D \cos\theta_B, \qquad (5.13)$$

where B is the width of the diffraction line measured at an intensity equal to half the maximum intensity (Fig. 5.19) (in radians), λ the X-ray wavelength (nm), D the crystallite size (nm), $2\theta_B$ the diffraction angle of the radiation corresponding to Bragg maximum, and K the shape factor, dependent on the symmetry (i.e., for the regular system slightly different from 1, we assume $K = 1$).

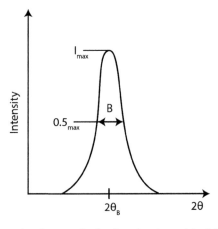

Figure 5.19 Determination method of grain size with Scherrer equation (B—full width at half maximum FWHM).

The shape of the crystallites affects the value of K, which can be 0.9 or 1.

The width of the diffraction reflexes also depends on factors related to apparatus used, for example: the diameter and width of the slots and a number of parameters such as the shape of the focus of X-ray tube, the distribution of intensity of the X-rays on the specimen surface, absorption of the sample, etc.

In order to avoid side effects of these parameters the method of internal standards is used. In this method, the substance of the crystallite size of about 1000 nm is added to the sample substance. The crystallite size is so great that the broadening of the diffraction reflexes of the standard substance is below the range of measurability.

A diffraction measurement of external standard and calculation of β_0 at FWHM for this standard can also be performed.

In Fig. 5.20, it can be seen that the smaller the crystallite size, the greater the broadening of reflection.

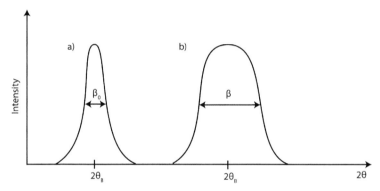

Figure 5.20 Half-width of diffraction reflex: (a) Shape reflex for the standard sample with the size of 10 μm; (b) shape reflex for the sample with the grain size below 100 nm.

Three different methods of determining the width B of the diffraction lines can be used.

$$B_1 = \beta - \beta_0, \qquad (5.14)$$

where β is the width of the diffraction line of the sample (measured at an intensity equal to half the maximum intensity) and β_0 is the width of the diffraction line of the standard (measured at an intensity equal to half the maximum intensity).

This method was proposed by Scherrer. It would be correct if the diffraction maxima had the shape of rectangles. It can be used when β_0 is several times smaller than β. Otherwise, when the difference $\beta - \beta_0$ is not small, the shape of the reflexes has to be taken into account. This shape is described most often by Gaussian and Lorentz function convolution (pseudo-Voigt). Then the following formula can be used:

$$B_2 = \sqrt{\beta^2 - \beta_0^2} \qquad (5.15)$$

Taking also into consideration statistical crystallite size distribution around the average value, the following formula can be used:

$$B_3 = \sqrt{(\beta - \beta_0)\beta^2 - \beta_0^2} \qquad (5.16)$$

Using Eqs. (5.15) and (5.16) is justified only in strict measurements.

A qualitative method to assess whether the crystallite size is greater than 100 nm is based on the effect of diffraction line broadening along with the grain refinement and a noticeable "split of $\alpha_1 \alpha_2$ doublet" of CuK$_\alpha$ line.

5.4.4 X-Ray Diffraction Methods

X-ray powder diffraction was introduced independently by Debye and Scherrer (in 1915 in Germany) and by Hull (in 1916 in the United States). Initially these methods relied on photographic film on which the angles and intensities of diffracted beams were recorded.

Debye–Scherrer powder camera is the simplest device for the measurement of a powder diffraction diagram. It consists of a cylindrical camera body with an entrance pinhole collimator and an exit beam collimator lying along the diameter of the camera (Fig. 5.21). The specimen is mounted as a thin cylinder at the central axis of the camera. A piece of film is placed inside the cylindrical wall of the camera and small holes are cut in the film for the entrance and exit collimators. A beam of X-rays is directed onto the specimen through the entrance collimator and the diffracted X-rays fall onto the film. The film after being developed is ready for the interpretation.

The application of Debye–Scherrer camera is the analysis of very small specimens. Acceptable patterns can be obtained from only a few milligrams of material. The camera technique has the disadvantage that it may be rather slow and exposure times are typically in the range of 1 to 6 h. In addition to the Debye–

Scherrer powder camera there are also available more specialized cameras for various applications.

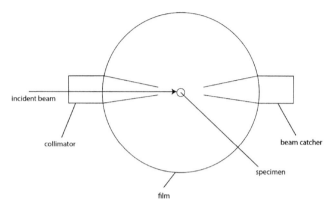

Figure 5.21 Debye–Scherrer camera.

A diffractometer is designed alike the Debye–Scherrer camera, but the main difference is the way of measuring of the intensity of a diffracted beam. The strip of film is replaced by a movable detector. There are many types of X-ray detectors (see Chapter 5.3.5) but they all convert incoming X-rays into pulses of electric current which is then processed by electronic components, i.e., computers.

The basic schematic of a diffractometer is shown in Fig. 5.22.

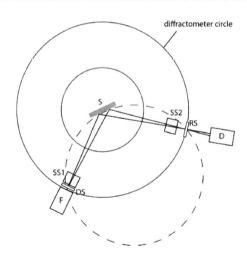

Figure 5.22 Schematic of an X-ray diffractometer.

This geometric arrangement is known as the Bragg–Brentano focusing system. A diverging beam from a line source F (i.e., X-ray tube) is falling onto the specimen S (it may be powder or bulk polycrystalline material), being diffracted and passing through a receiving slit RS to the detector D. Distances FS and SRS are equal.

The amount of divergence is determined by the effective focal width of the source and the aperture of the divergence slit DS. Axial divergence is controlled by two sets of parallel plate collimators (called Soller slits) SS1 and SS2 placed between focus and specimen, and specimen and receiving slit respectively. The use of the narrower divergence slit gives a smaller specimen coverage at a given diffraction angle, thus allowing the attainment of lower diffraction angles where the specimen has a larger apparent surface (thus larger values of d are available). This is achieved, however, only at the expense of intensity loss. A scintillation detector is typically placed behind the scatter slit, and this converts the diffracted X-ray photons into voltage pulses. By synchronizing the scanning speed of the goniometer with the recorder, a plot is obtained of 2θ° against intensity, called the diffractogram. A diffracted beam monochromator may also be used in order to improve signal-to-noise characteristics.

References

1. Kittel, C. (2004). *Introduction to Solid State Physics*, Wiley, pp. 3–6.
2. Stachurski, Z. H. (2011). On structure and properties of amorphous materials, *Materials*, **4**, 1564–1598.
3. Simon, S. H. (2013). *Oxford Solid State Basics*, Oxford, pp. 118–124.
4. Miller, W. H. (1839). *A Treatise on Crystallography*, Cambridge, pp. 1–4.
5. Weertman, J., and Weertman, J. R. (1992). *Elementary Dislocation Theory*, Oxford University Press.
6. Lejcek, P. (2010). *Grain Boundary Segregation in Metals*, Springer-Verlag Berlin Heidelberg.
7. Röntgen, W. H. (1898). Über eine neue Art von Strahlen, *Annalen der Physik* **300**: 1–11.
8. McQuarrie, D. A. (2008). *Physical Chemistry*. 2nd ed., United States of America: University Science Books.

9. Sproull, W. T. (1946). *X-Rays in Practice*, McGraw-Hill.
10. Knoll, G. F. (1999). *Radiation Detection and Measurement*, 3rd ed., Chapter 11, John Wiley & Sons.
11. Bragg, W. L. (1913). The diffraction of short electromagnetic waves by a crystal, *Proceedings of the Cambridge Philosophical Society*, **17**, 43–57.
12. Friedrich, W., Knipping, P., and Laue, M. (1913). Interferenzerscheinungen bei Röntgenstrahlen, *Annalen der Physik* **346**: 971–988.
13. Scherrer, P. (1918). *Göttinger Nachrichten Gesell*, vol. 2, p. 98.

Chapter 6

Atomic Force Microscopy in Hydrogen Storage Materials Research

Jaroslaw Jakubowicz

*Institute of Materials Science and Engineering,
Poznan University of Technology, Poznan, Poland*

jaroslaw.jakubowicz@put.poznan.pl

6.1 Principles of the AFM Technique

Atomic force microscopy (AFM) is a powerful technique that was developed about 30 years ago [1]. The AFM is a part of the microscopy technique family, called scanning probe microscopy (SPM) [2]. This microscopic family includes the most important scanning tunneling microscope (STM), AFM, magnetic force microscope (MFM), and near field scanning optical microscope (NFSOM) [1–3]. The principle of all SPM microscopes is the same: The mechanical probe scans the surface, and the generated signal (different in different techniques) is converted into topography or physical properties map distribution image. The main advantage of the SPM is its very high resolution and visualization of 3D surface topography. The resolution is in the range of 0.1–10 nm in the x–y plane and up to 0.01 nm in the z axis. The

---

Handbook of Nanomaterials for Hydrogen Storage
Edited by Mieczyslaw Jurczyk
Copyright © 2018 Pan Stanford Publishing Pte. Ltd.
ISBN 978-981-4745-66-6 (Hardcover), 978-1-315-36444-5 (eBook)
www.panstanford.com

high resolution is due to the construction of the microscope, composed of the piezoelectric scanner (which moves the tip or sample in the very precise way) and the sharp tip, which scans the surface (Fig. 6.1a). In all SPM techniques, the tip is moved in the raster way, probing the surface in up to few thousands lines and the data is measured in each line up to a few thousand points.

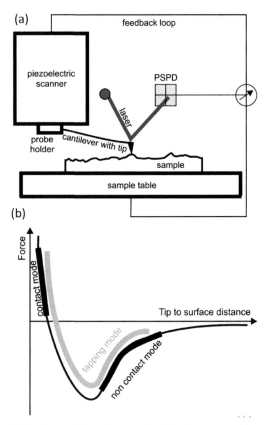

Figure 6.1 AFM scheme (a) and van der Waals curve with marked AFM working modes (b).

The AFM has many advantages over the other SPM techniques, including research of the all materials type, such as conducting, nonconducting, hard and soft, as well as simplicity of operation, operating in different environments, including ambient, vacuum, liquid, or any gaseous atmosphere [4].

In the AFM, the scanning procedure is based on the measurement of the force, which acts between tip atoms and surface atoms (Fig. 6.1b). In the AFM, many atoms of the tip act with many atoms of the investigated surface; thus, the resulting surface image is a result of the averaging of the force acting between millions of interacting atoms.

When the distance between tip atoms and surface atoms is relatively long, the attractive force acts between them, and the AFM works in non-contact mode (cantilever vibrates over the surface in the range of up to tens of nanometers, without touching the surface). When the distance between the interacting atoms is close enough, the repulsive force strongly increases and the AFM works in the contact mode (no cantilever vibrations are induced) and the tip runs over the surface in close proximity <1 nm. While the cantilever vibrates and the tip touches the surface in the tapping way, the tip is alternately in the range of attractive and repulsive force and the method is called the tapping mode. The contact mode is suitable for hard materials, because there is very high possibility of surface failure during the tip-surface friction. The non-contact mode is suitable for soft materials. The absence of the tip–surface contact gives a non-damaged surface; however, it is possible to visualize some liquid monolayer drop, which lies on the surface, instead of a real surface (during scanning the liquid drop is not penetrated by the tip, hence it is shown on the image). Considering the disadvantages of the contact and non-contact scanning procedures, the best results in most cases will be achieved using the tapping mode. In this method, the tip has a local point contact with the surface (the tip is not pulled along the surface) during cantilever vibrations allowing minimal surface damage. If the liquid drop lies on the surface (for example, during scanning in ambient), then the drop is penetrated by the tip, resulting in true surface image formation.

The AFM equipment has a few characteristic parts, including piezoelectric scanner (made from PZT), which is necessary for precise movement with subatomic resolution. Due to different constructions of the AFM, the scanner can move the probe (Fig. 6.1a) or move the sample table. The probe mounted to the scanner allows scanning the bigger sample. In other cases, if the sample table is mounted on the scanner, the sample has very limited size (for example, to 10 mm side size). In the probe holder, the AFM probe is mounted. In the case of the Quesant microscope (Fig. 6.2), the probe is composed of a pre-mounted typical probe (for example, Nanosensors™) [5] fixed to the probe grip (Fig. 6.3), cantilever, and tip at end of cantilever. The pre-mounted probe is provided by NanoAndMore™, for example [6]. While the probe moves over the surface, the cantilever is bent because of the force acting between the tip and the sample surface atoms. To transform the cantilever bending on the surface image, the optical method is commonly used. The laser is directed toward the end of the cantilever and after reflection is guided into the position-sensitive photo detector (PSPD) (Fig. 6.1). The laser position changes on the PSPD allow to detect changes in the surface topography (the signal from PSPD is transformed into the surface image). For proper working, the AFM and all SPMs work in a feedback loop to give the best-quality image as well as to avoid probe breaking.

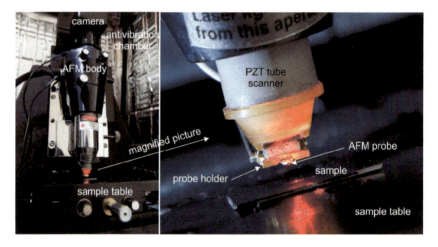

Figure 6.2 Quesant Q-Scope 250 AFM in antivibration chamber.

For best results, the SPM should be kept on the lowest level of a building as well as in the antivibrational table and/or chamber to avoid and limit excessive vibrations affecting the tip–sample surface interactions.

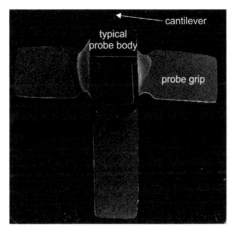

Figure 6.3 Premounted NanoAndMore™ Nanosensors™ AFM probes.

6.2 Measurement Procedure

The AFM can be used to scan many materials in different forms, for example nanopowders or solid surfaces. No special procedure is requested for sample preparation; however, the surface roughness is limited to a few micrometers, depending on the PZT scanner elongation (longer scanner piezoelectric tube results in its longer elongation as well as larger scan size). Thus, in some cases, a metallographic sample preparation is necessary (grinded, polished, and/or etched). For a solid surface, depending on its properties, any scanning mode can be useful. When the nanopowders are investigated, the main problem is related with sample (powders) fixing and scanning mode choice, to avoid the movement of the powders. The best way for nanopowders scanning is to choose the less invasive mode, i.e., non-contact or tapping mode, which limits the physical contact (and force) with investigated powders. The second important factor is to fix the powder using double-sided glue tape put on the metal pad. On the surface, then the powder is poured in small quantities, which should be pressed and the excess of loose powder is blown out. In this way, the powder sample is prepared for scanning. During the scanning, it is important to choose a supersharp tip (Fig. 6.4) to limit the volume of the tip interacting with powders, which gives the best image resolution and low scan speed (for example, 2–3 Hz) and, furthermore, limits the powder moving through the moving tip. During scanning, the cantilever vibrates with frequency close to resonant frequency (at about 80% resonant frequency value).

The SuperSharpSilicon™ probe (SSS-NCL), provided by Nanosensors™ (Fig. 6.4, Table 6.1), has good properties for nanopowders investigation. The probes are designed for the non-contact mode or the tapping mode [5]. This AFM probe combines high operation stability with outstanding sensitivity and fast scanning ability [5]. The probe of a monolithic material has a tip radius of curvature <5 nm, typical aspect ratio at 200 nm from tip apex in the order of 4:1, half cone angle at 200 nm from tape $x < 10°$, highly doped to dissipate static charge, chemically inert [5]. The probe has reflex coating of about 30 nm-thick aluminum on the detector side of the cantilever, which enhances the reflectivity of the laser beam by a factor of about 2.5 [6]. The coating prevents light from interfering within the cantilever. Additionally, a single-wall nanotube can be mounted to the tip for enhancing the resolution and the aspect ratio [7].

Scanning of the relatively low area limited the powder movement, providing a good-quality image. For example, the scan size in the first approach usually should not exceed 2 × 2 μm; however, in some cases a larger scan size is possible to realize without powder movement.

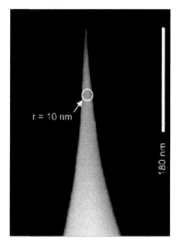

Figure 6.4 SuperSharpSilicon™ AFM probe [5].

Table 6.1 Cantilever SuperSharpSilicon™ data [5]

Property	Nominal value	Specified Range
Thickness (μm)	7	6.0–8.0
Mean width (μm)	38	30–45
Length (μm)	225	215–235
Force Constant (N/m)	48	21–98
Resonance frequency (kHz)	190	146–196

6.3 Hydrogen Storage Nanomaterial Imaging

In the recent years, hydrogen storage nanomaterials have been extensively investigated [8–12]. Among different microscopic techniques, the AFM has a great potential for nanomaterial imaging.

Mechanical alloying is a common way of producing new hydrogen storage nanomaterials, used by the Jurczyk group [2, 9–12]. In this process, the mechanical milling of elemental powders in stoichiometric or non-stoichiometric composition leads to the formation of alloys or composites in the form of powders, i.e., formation of agglomerates composed of nanograins. The TEM is a main type of microscope intended for revealing nanostructure and structure features. However, in the past years, the AFM was successfully applied for the characterization of mechanically alloyed powders, too [2, 9, 13].

The example images of hydrogen storage nanomaterials prepared by mechanical alloying are shown in Figs 6.5–6.8. Different systems were investigated, for example TiNi, TiNi+Pd, TiNi+MWCNTs, Ti_2Ni, Ti_2Ni+Pd, Ti_2Ni+MWCNTs, and Ti_2Ni+Pd+MCNTs [2, 9].

The data recorded by AFM can be presented in a different form, taking into account different signals (Fig. 6.5), i.e., as the typical height image 3D and 2D (a, b), error image (c) taking the feedback error signal, which contains some additional information about the topography for more precise recovery of the relief [14], and the phase image (d) which detect surface structure or elasticity property. The last two modes of result presentation give more details and more contrast in comparison to the height mode.

The AFM image presented in Fig. 6.6 shows mechanically alloyed (through 8 h) TiNi mixed through 1 min with 5 wt% MWCNTs. Carbon nanotubes are clearly visible—the lines pointed by arrows. After a short period of intensive milling, MWCNTs are uniformly distributed in the TiNi matrix. Based on the AFM pictures, the average grain size was estimated to be 65 nm. About 70% of the grains had size below 100 nm. The smallest observed grains had 30 nm size, but some bigger ones were observed too, with size up to 240 nm. After ball milling with 5 wt% MWCNTs, the particle size of the TiNi alloy decreased explicitly [2]. Due to this microstructure, the high catalytic dispersion and activity of MWCNTs prevent as-milled alloys from adhering effectively and stabilize smaller powder particles. To conclude, the AFM is a powerful tool for such type of composite nanostructure visualization.

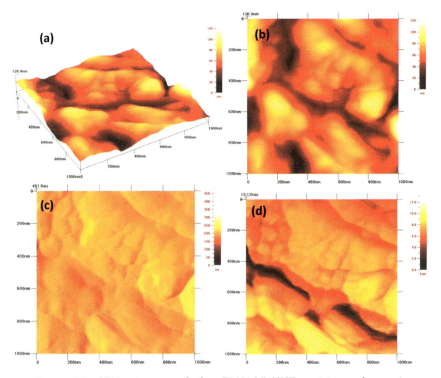

Figure 6.5 AFM pictures of the Ti_2Ni+MWCNTs mixture shown in different modes: height 3D (a), height 2D (b), error (c), and phase picture (d); in all cases, the same area is presented.

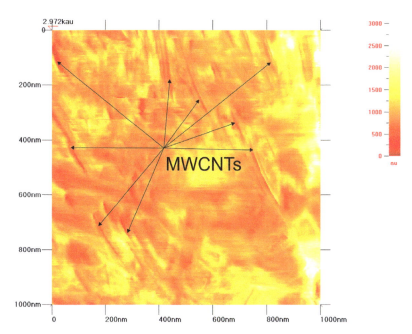

Figure 6.6 Carbon multiwall nanotubes in the TiNi matrix; error mode picture.

The phase mode picture type reveals more clearly the nanostructure compared with the standard height mode picture (Fig. 6.7). The contrast is higher and the grains boundaries are significantly enhanced in the phase mode presentation (the data for the phase mode are recorded simultaneously with the height mode data during the tapping procedure scanning).

The picture recorded by AFM can be useful not only for topography characterization (Fig. 6.8a), but for the quantitative analysis (Fig. 6.8b), too. For example, if the picture shows the grains, then their size can be measured and the grain size distribution can be presented in a graphical form (Fig. 6.8b).

Not only nanomaterials in the form of nanopowders may be investigated. For example, mechanical pressing of the powders lead to the formation of green compacts, where the grains are mechanically fixed between each other, providing their relatively strong mechanical strength. The sample can be relatively quite easy scanned by the AFM probe (without using a glue tape for the powders).

For real powders' physicochemical bonding, sintering is necessary. In the case of nanopowders, the best results give hot pressing at lower temperature, shorter time and pressure in comparison to the conventional powder metallurgy process. In that form, nanomaterials can be investigated successfully; however, grain growth should be taken into account in comparison with the starting nanopowders.

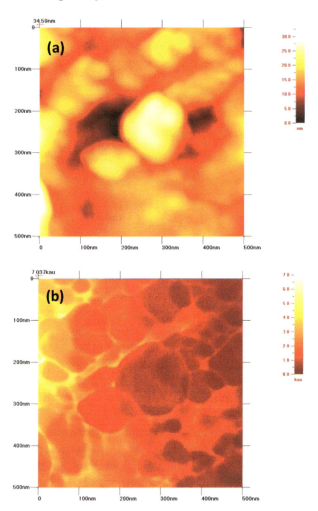

Figure 6.7 AFM pictures of Ti$_2$Ni alloy; height 2D (a) and error mode image (b).

Hydrogen Storage Nanomaterial Imaging | 113

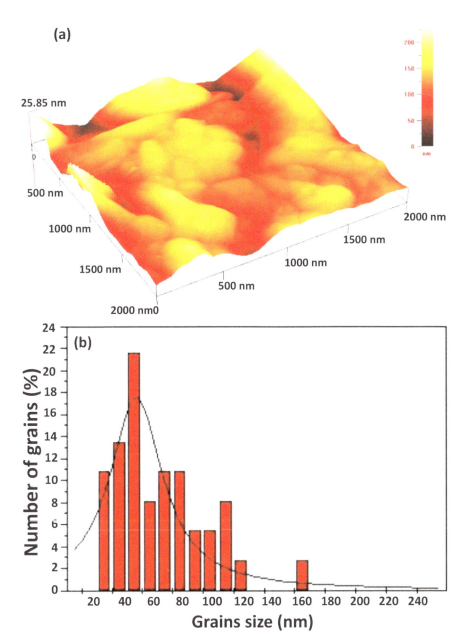

Figure 6.8 AFM picture of the TiNi+5%Pd+5%MWCNT mixture (a) and the grain size distribution (b).

6.4 Typical Problems Observed during Nanomaterial Imaging

The AFM is an excellent tool for material research; however, it is possible to find some artifacts on the recorded pictures.

If the tip becomes worn-out or the debris attaches itself to the end of the tip, the features in the image may all have the same repeatable shape [2]. The image shows not the morphology of the surface features, but the debris, for example. The artifact is sometimes called "tip imaging" (Fig. 6.9) and is often observed when using old, contaminated, or low-quality tip.

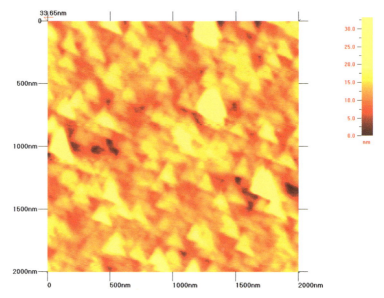

Figure 6.9 Typical "tip imaging" error picture often recorded during data acquisition.

When the powder is not strongly bonded to the glue, there is a possibility of the tip pulling the powder on the surface. The obtained image is blurred (Fig. 6.10). The phenomenon is usually observed while scanning a relatively large area, when the possibility of powder movement increases. To avoid the powder movement, a smaller area should be scanned or the loose powder should be removed from the surface or the non-contact mode should be applied instead of the tapping mode.

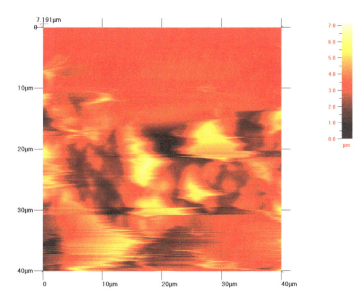

Figure 6.10 Typical effect of the powders moving under the tip during scanning procedure.

Another artifact or shortcoming may be the scanning not fully useful area. For example, as the result, a part of the image shows nanopowders, whereas the other part is a metal plate covered by the glue (Fig. 6.11). However, sometimes this border presentation may be useful when two phases or two materials stick together.

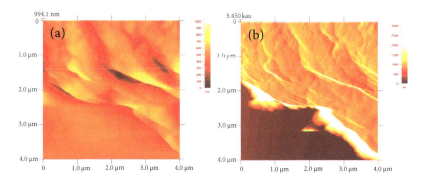

Figure 6.11 AFM picture of the $Ti_2Ni+5\%Pd+5\%MWCNT$ mixture showing boundary of the powder agglomerate; the darker area is a metal plate on which the powder lies; height 2D (a) and error mode image (b).

6.5 Conclusion and Future Perspectives

The AFM is a useful tool for the investigation of hydrogen storage nanomaterials in the form of nanopowders, providing slightly less data than high-resolution transmission microscope (HRTEM). Usually the powder size, morphology, phases, and grain size are revealed in standard procedure investigations.

Recent research done by Sugimoto et al. [15] shows that the AFM can be successfully applied to atoms moving at room temperature, as well as in the atomic plane (not only ad-atoms as in the case of scanning tunneling microscope). The procedure may be applied in future for the creation of new compounds for hydrogen storage batteries and can also be applied for the study of the position of the given atoms in a structure. The force is measured between two adjacent tip and surface atoms (the force is different for different combination of atomic pairs)—this is the way for atoms identification using AFM. The results can be easily shown on the surface image by the correlation of specific interatomic force and color assigned for the given atom type.

Although in this chapter, the AFM technique was presented, most SPM techniques should be useful for hydrogen storage nanomaterial study.

References

1. Binnig, G., Quate, C. F., and Gerber, C. H. (1986). Atomic force microscope. *Phys. Rev. Lett.,* **56**: 930–933.
2. A practical guide to scanning probe microscopy, www.veeco.com.
3. Mironov, V. L. (2004). *Fundamentals of the Scanning Probe Microscopy*. Russian Academy of Sciences Institute of Physics of Microstructures, Nizhniy Novgorod.
4. Meyer, E., Hug, H. J., and Bennewitz, R. (2004). *Scanning Probe Microscopy, The Lab on a Tip*, Springer-Verlag Berlin Heidelberg, 215p.
5. www.nanosensors.com (Nanosensors Company).
6. www.nanoandmore.com (Nano and More Company).
7. Stevens, R. M. (2009). New carbon nanotube AFM probe technology. *Mater. Today,* **12**: 42–45.

8. Balcerzak, M., Nowak, M., Jakubowicz, J., and Jurczyk, M. (2014). Electrochemical behavior of nanocrystalline TiNi doped by MWCNTs and Pd. *Renew. Energy*, **62**: 432–438.
9. Balcerzak, M., Jakubowicz, J., Kachlicki, T., and Jurczyk, M. (2015). Hydrogenation properties of nanostructured Ti$_2$Ni-based alloys and nanocomposites. *J. Power Sources,* **280**: 435–445.
10. Gasiorowski, A., Iwasieczko, W., Skoryna, D., Drulis, H., and Jurczyk, M. (2004). Hydriding properties of nanocrystalline Mg$_{2-x}$M$_x$Ni alloys synthesized by mechanical alloying (M = Mn, Al). *J. Alloys Comp.*, **364**: 283–288.
11. Jurczyk, M., Jankowska, E., Nowak, M., and Jakubowicz, J. (2002). Nanocrystalline titanium-type metal hydride electrodes prepared by mechanical alloying. *J. Alloys Comp.*, **336**: 265–269.
12. Majchrzycki, W., and Jurczyk, M. (2001). Electrode characteristics of nanocrystalline (Zr, Ti)(V, Cr, Ni)$_2$ compound. *J. Power Sources,* **93**: 77–81.
13. Jakubowicz, J. (2003). Application of atomic force microscopy in microstructure analysis of mechanically alloyed Nd$_2$Fe$_{14}$B/α-Fe-type nanocomposites. *J. Alloys Comp.* **351**: 196–201.
14. www.ntmdt.com (NT-MDT Company).
15. Sugimoto, Y., Pou, P., Custance, O., Jelinek, P., Abe, M., Pérez, R., and Morita, S. (2008). Complex patterning by vertical interchange atom manipulation using atomic force microscopy. *Science*, **322**: 413–417.

Chapter 7

Characterization of Hydrogen Absorption/Desorption in Metal Hydrides

Mateusz Balcerzak

Institute of Materials Science and Engineering,
Poznan University of Technology, Poznan, Poland

mateusz.balcerzak@put.poznan.pl

7.1 What Is a Sievert-Type Apparatus

In order to obtain knowledge about hydrogen storage properties (sorption from gas) of material, thermodynamic measurements have to be done. There are a number of methods to determine pressure–composition–temperature (PCT) curves. One of them is a volumetric method that uses a Sievert-type apparatus. This device most often is equipped with a calibrated and thermostated volumes and pressure gauges. An example of the apparatus is presented in Fig. 7.1. Inter alia, the advantages of this method are high accuracy of measurements (depends on the accuracy of the pressure gauges) and the ability to test a wide range of pressure equilibrium. The Sievert-type apparatus is a device that measures a variation of gas (in this case H_2) pressure in a

---

Handbook of Nanomaterials for Hydrogen Storage
Edited by Mieczyslaw Jurczyk
Copyright © 2018 Pan Stanford Publishing Pte. Ltd.
ISBN 978-981-4745-66-6 (Hardcover), 978-1-315-36444-5 (eBook)
www.panstanford.com

reaction chamber. Pressure changes correspond to the changes in hydrogen concentration in the tested material at a given temperature.

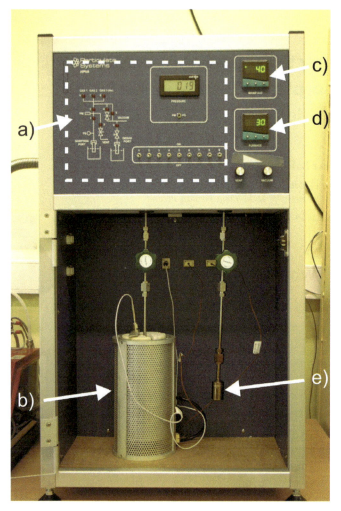

Figure 7.1 Sievert-type apparatus: (a) scheme of apparatus with valves switches and display of actual pressure in the measuring chamber, (b) furnace, (c) manifold temperature control panel, (d) sample temperature control panel, and (e) measuring chamber.

A typical pressure–concentration–temperature plot is shown in Fig. 7.2. The first area (marked as I) corresponds to the initial dissolving of hydrogen in material, resulting in the creation of a solid solution of hydrogen in a metal matrix α-phase. With increasing of hydrogen pressure and hydrogen concentration in the metal, nucleation and growth of new metal hydride β-phasee is observed. Thee second area (marked as II), which is the plateau area, corresponds to a mixture of a solid solution α-phase and metal hydride β-phase. This area is the most important with respect to hydrogen storage properties because the length of the plateau provides information on how much H_2 can be stored reversibly without big changes in pressure. In the last area (marked as III), the α-phase is completely converted to the β-phase [1].

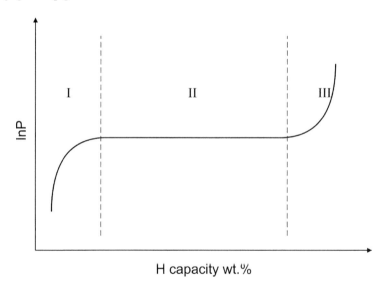

Figure 7.2 Schematic pressure–concentration–temperature plot: (I) α-phase area, (II) $\alpha + \beta$ phases mixture area, (III) β-phase area.

7.2 Preparation of Material to PCT Tests

For all studies involving the use of hydrogen, it is very important to ensure safety because of the highly explosive nature of this gas.

Before each PCT measurement, a material should be weighed. Most of commercially produced apparatuses automatically calculate the hydrogen concentration by changes in hydrogen pressure and material weight.

Materials that are often studied by using Sievert-type apparatus are alloys, which can be easily oxidized. According to the method of preparation and storage of the material, they may be less or more oxidized. In the context of hydrogen sorption, oxides are undesirable because of their negative impact on hydrogen storage properties. The presence of the oxide film on the studied material surface causes inactive state for hydrogen sorption. For this reason, each of material should be degassed before PCT measurements. Most of the systems for this purpose consist of a vacuum pump and a furnace. The degassing temperature should be selected for

- stability of oxide phase (the temperature should be high enough to allow for the desorption of the oxides);
- diffusion (the higher the temperature, the faster the diffusion process between the test material and the material of the measuring chamber; in the case of multi-phase materials, diffusion of elements between the various phases is also possible);
- studying material grain growth (progressing with increasing temperature; together with the growth of grains, surface area is decreasing, which affects the worsening of sorption properties of the material).

In the next step, fully degassed materials should be subjected to activation process. This process is based on initial hydrogenation of sample under high pressure and temperature. Activation is intended to obtain maximum H-capacity and faster hydrogenation and dehydrogenation kinetics. This process depends on the surface structure and barriers. Initial hydrogenation causes the fracturing of the material, which results in the formation of smaller particles increasing the reaction surface area. The quantity of activation samples depends on the type of material [2].

Figure 7.3 presents the absorption curve of the studied material before and after activation process. As shown in the figure, the activation process highly increases the plateau region and hydrogen capacity.

Activated samples should be once again degassed before proper measurement.

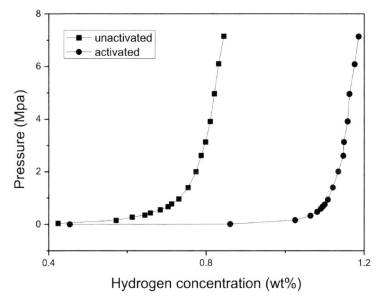

Figure 7.3 PCI absorption curves measured on inactivated and activated material.

7.3 Types of Tests That Can Be Performed Using Sievert-Type Apparatus

Sievert-type apparatus can be used to perform different types of measurements which can provide different types of information [1, 3].

PCT curves are the usually received results from Sievert-type apparatus. It is important to correctly choose the measuring pressure to not miss important points on the curve of sorption and desorption. Moreover, it is also important to determine the condition for receiving the equilibrium pressures.

PCT curves can provide answers to the following questions:

- Does material actively absorb at a given temperature?
- What is the pressure of plateau?
- What is the plateau region and is it flat or sloping?
- Does the hysteresis between sorption and desorption curves appear?

- How much of hydrogen can be desorbed from the material at a given temperature?
- What is the reversibility of hydriding-dehydriding process?

Moreover, PCT curves obtained at different temperatures can be used to make Van't Hoff plot. From this plot, information about thermodynamic properties of material can be obtained.

Sometimes it is worthwhile to perform a few PCT tests on the same material in order to obtain the value of maximum sorption capacity as a function of sorption and desorption cycles. This kind of measurements can provide information about the activation properties of each material (in which cycle the maximum capacity is reached) and about resistance to damage caused by gaseous impurity (which lower the capacity).

Sievert-type apparatus can be also used to test the kinetics of the hydrogenation process of a material. This measurement shows how the pressure of hydrogen in the reaction chamber changes with time. It can give information about the presence or absence of the absorption incubation time. Materials with good kinetics absorb without any incubation time. Moreover, the time of obtaining equilibrium pressure when the material has absorbed maximum hydrogen is also important. A material with potential commercial use should absorb hydrogen as quick as possible.

XRD studies can be used to verify changes between the structure of a starting material and the material after hydrogenation and dehydrogenation tests. It can help in finding a correlation of PCI measurements with structure of studied material.

References

1. Varin, R. A., Czuko, T., and Wronski Z. S. (2009). *Nanomaterials for Solid State Hydrogen Storage*, Springer Science + Business Media, Chapter 1.4, pp. 56–74.

2. Balcerzak, M., and Jurczyk, M. (2015). Influence of gaseous activation on hydrogen sorption properties of TiNi and Ti_2Ni alloys. *J. Mater. Eng. Perform.* **24**: 1710–1717.

3. Balcerzak, M., Jakubowicz, J., Kachlicki, T., and Jurczyk, M. (2015). Hydrogenation properties of nanostuctured Ti_2Ni-based alloys and nanocomposites. *J. Power Sources* **280**: 435–445.

Chapter 8

Electrochemical Characterization of Metal Hydride Electrode Materials

Mateusz Balcerzak and Marek Nowak

*Institute of Materials Science and Engineering,
Poznan University of Technology, Poznan, Poland*

mateusz.balcerzak@put.poznan.pl, marek.nowak@put.poznan.pl

8.1 Fundamentals of Electrochemical Research

Every material for potential use as a negative electrode material in Ni-MH$_x$ rechargeable batteries must be first electrochemically tested. In the past, a large number of materials with different composition were tested for their electrochemical properties. There are many alloy families that can be investigated as negative electrodes for Ni-MH$_x$ batteries: AB, A$_2$B, AB$_2$, and AB$_5$. A represents a metallic element with strong affinity for hydrogen and B a metallic element with a weak affinity for hydrogen [1–5]. Most often, these kind of studies are conducted on a battery testing system. Figure 8.1a shows an example of measuring device—Multi-channel Battery Interface ATLAS 0461.

Electrochemical measurements are often done using three-compartment glass cell (H-shape)—shown in Fig. 8.1b. The measuring set consists of tested electrode, reference electrode

---

Handbook of Nanomaterials for Hydrogen Storage
Edited by Mieczyslaw Jurczyk
Copyright © 2018 Pan Stanford Publishing Pte. Ltd.
ISBN 978-981-4745-66-6 (Hardcover), 978-1-315-36444-5 (eBook)
www.panstanford.com

and counter electrode. For these studies NiOOH/Ni(OH)$_2$ counter electrode is used. The reference electrode can be for example mercury oxide electrode (Hg/HgO/6 M KOH). Before each measurement, it is important to check the connection between every electrodes and battery testing system.

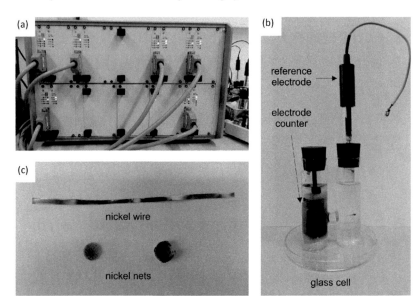

Figure 8.1 An example of battery testing system (a), H-type glass cell with reference and counter electrodes (b), and nickel wire and nets for negative electrode production (c).

The basis of energy storage process in Ni-MH$_x$ rechargeable batteries is the redox reaction. During discharging on negative and positive electrodes, oxidation and reduction reactions occur respectively. The reaction scheme is as follows:

MeH + OH$^-$ → Me + H$_2$O + e$^-$

NiOOH + H$_2$O + e$^-$ → Ni(OH)$_2$ + OH$^-$

For this kind of measurements, 6 M KOH is used as an electrolyte. KOH is an ion conductor and does not participate in the chemical reaction occurring within the cell. Sometimes, instead of KOH, NaOH is used. However, NaOH weakly conducts ions.

8.2 Preparation of Materials for Electrochemical Measurements

Before measurement, material which is most often in powder form should be mixed with nickel (in various proportions). Addition of nickel to electrode material increases electrical conductivity, catalyzes electrode reaction and surface dissociation, prevents the diffusion of oxygen to the internal part of the alloy, and increases a cycle life of electrodes [6]. Some groups prepare electrode according to the "latex" technology using addition of black carbon to increase electrode conductivity and polytetrafluoroethylene as a binder [1]. Moreover, Barsellini et al. used teflonized carbon in weight ration 1:1 with electrode material [2].

In the following step, a mixture of powders should be cold pressed between two nickel nets (Fig. 8.1c) to form a round-disc. Nickel nets act as the current collector. After pressing, a Ni wire should be attached to the Ni net by spot-welding.

It is important to keep the samples away from air to reduce the adverse effect of oxidation—electrodes can be partially produced in a glove box filled with argon. However, almost all of materials need to be activated. This process is carried out to eliminate the adverse effect of oxidation. There are various methods of the activation of materials before electrochemical measurements, one of which is based on soaking of electrodes in electrolyte for 24 h at room temperature with additional etching at 100°C for 1 h in the same solution [6].

8.3 The Results of Electrochemical Measurements

An example of electrochemical result has been presented in Section 10.2.1, titled "Cycle-Life curves" (Chapter 10). Two most important factors that determine the suitability for negative electrode for Ni-MH$_x$ rechargeable batteries is maximum discharge capacity and cycle stability. Materials with potential commercial use should be characterized by high discharge capacity and good cycle stability. Cycle stability can be evaluated by a capacity retaining rate: $R_h = (C_n/C_{max}) \times 100\%$, where C_n and C_{max} are discharge capacities at the n-th cycle and maximum discharge capacity, respectively. Electrochemical measurements can also give information about activation properties of material—showing how many cycles of charging and discharging are needed to obtain maximum discharge capacity.

The obtained experimental values depend on many factors. The most important one is the material: its structure, phase composition, chemical composition, size of crystallites, etc. Moreover, these values also depend on used amperage on gram of used material and cut-off potential vs. Hg/HgO/ 6 M KOH. Examples of amperage and cut-off potential are given in Section 10.2.2, titled "Electrodes and Electrochemical Measurements Conditions" (Chapter 10). Moreover, the temperature of measurements hardly affects electrochemical properties of materials. Temperature measurements can give information about the potential application of the studied system.

It is important to note that by electrochemical measurements, it is possible to obtain PCT isotherms which describe the dependence of the hydrogen equilibrium pressure to the amount of hydrogen dissolved and/or incorporated into the solid phase at different constant temperatures. There is a thermodynamic correlation between the equilibrium pressure measured in the gas-phase reaction and the electrode potential measured in an electrochemical cell. Both techniques are proper for PCT isotherm measurements and there is good correlation between them. The disadvantage of using electrochemical devices for obtaining PCT curves is that it can be applied only for metal hydride alloys that can be electrochemically charged/discharged [5].

Moreover, using van't Hoff plots, it is possible to calculate enthalpy and the entropy of the hydride formation/decomposition process from variation of the logarithm of the equilibrium pressure with temperature [5].

References

1. Ayari, M., Ghodbane, O., and Abdellaoui, M. (2015). Elaboration and electrochemical characterization of $LaTi_2Cr_4Ni_5$-based metal hydride alloys. *Int. J. Hydrogen Energy* **40**: 10934–10942.

2. Barsellini, D., Visintin, A., Triaca, W. E., and Soriaga, M. P. (2003). Electrochemical characterization of a hydride-forming metal alloy surface-modified with palladium, *J. Power Sources* **124**: 309–313.
3. Achyuthal Babu, R. S., Viswanathan, B., and Srinivasamurthy, S. (2000). Electrochemical characterization of LaNi$_5$ hydride electrode, *Indian J. Chem.* **39A**: 690–696.
4. Kleperis, J., Wojcik, G., Czerwinski, A., Skowronski, J., Kopczyk, M., and Beltowska-brzezinska, M. (2001). Electrochemical behavior of metal hydrides, *J. Solid State Electrochem.* **5**: 229–249.
5. Bliznakov, S., Lefterova, E., Bozukov, L., Popov, A., and Andreev, P. (2004). Techniques for characterization of hydrogen absorption/desorption in metal hydride alloys, *Proceedings of the International Workshop "Advanced Techniques for Energy Sources Investigation and Testing"* **P19**: 1–6.
6. Balcerzak, M. (2016). Electrochemical and structural studies on Ti-Zr-Ni and Ti-Zr-Ni-Pd alloys and composites. *J. Alloys Comp.* **658**: 576–587.

Chapter 9

TiFe-Based Hydrogen Storage Alloys

Marek Nowak and Mieczyslaw Jurczyk

*Institute of Materials Science and Engineering,
Poznan University of Technology, Poznan, Poland*

marek.nowak@put.poznan.pl, mieczysław.jurczyk@put.poznan.pl

9.1 Phase Diagram and Structure

Figure 9.1 shows the Ti–Fe binary phase diagram [1]. Titanium and iron form two stable intermetallic compounds: TiFe and TiFe$_2$ (Table 9.1). TiFe was discovered as a metal hydride by Reilly and Wiswall [2]. This alloy has been considered to be a promising hydrogen storage material due to its excellent hydrogen storage properties. The total hydrogen capacity is 1.9 wt%. TiFe crystallizes in the bcc structure. The cell parameter of this phase equals a = 2.973 Å [3].

The TiFe alloy has good thermodynamic properties, good H-capacities, and low raw material cost. But there are some problems with activation, gaseous impurities, and upper plateau instabilities. The activation process is necessary because of the presence of an oxide film which obstructs hydrogen absorption. In the case of TiFe alloy, the activation of this alloy involves heating to a temperature of about 450°C in vacuum and annealing

---

Handbook of Nanomaterials for Hydrogen Storage
Edited by Mieczyslaw Jurczyk
Copyright © 2018 Pan Stanford Publishing Pte. Ltd.
ISBN 978-981-4745-66-6 (Hardcover), 978-1-315-36444-5 (eBook)
www.panstanford.com

in hydrogen at the pressure about 7 bars, followed by cooling down to room temperature and admission of hydrogen at a pressure of 35–65 bars [4, 5]. This alloy seems to be the one used in the hydrogen tanks of fuel cell–powered submarines.

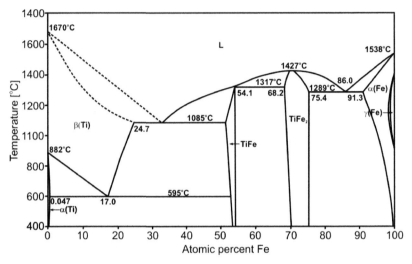

Figure 9.1 Phase diagram of Ti–Fe.

Table 9.1 Phases in Ti–Fe system

Phase	Iron content [% Fe]	Crystal structure
α Ti	0–0.047	hP2
β Ti	0–24.7	cI2
TiFe	51.3–54.1	cP2
TiFe$_2$	68.2–75.4	hP12
α Fe	91.3–100	cI2
β Fe	99.5–100	cF4
ω	(a)	hP3

(a) Metastable phase.

Work on ball-milled TiFe has shown that nanocrystalline alloys are easier to activate than conventional alloys [6, 7]. The final properties of this alloy after ball milling depend on the oxygen content. High O_2 content (>2.9 at%) leads to the formation of an amorphous TiFe phase [6]. If the oxygen content is larger than 5 at%, a Ti-rich amorphous phase is formed with imbedded unreacted Fe residual particles [8].

Study of nanocrystalline TiFe prepared by ball-milling has shown that this material tends to be oxidized easily. On the nanoparticle surface, the TiO_2, Fe_2TiO_5 and iron clusters are formed [9, 10]. Milling intensity is also an important factor for the achievement of a specific structure [11]: At low milling energy the intermetallic compound $Ti_{50}Fe_{50}$ is synthesized, while at higher milling intensities a partly amorphous material is formed.

Nanocrystalline TiFe has much higher hydrogen solubility at low pressure than conventional TiFe [9, 12]. The activation step has been shown to be unnecessary when TiFe is milled under argon with addition of a small amount of nickel or palladium [13, 14]. Crystalline nano-grains and highly disordered (amorphous) grain boundaries are visible after ball milling of TiFe [15].

In nanocrystalline TiFe with palladium catalyst, it was found that structural relaxation leads to an increase in the solubility gap and a change in the slope of the plateau in the PCI curve [16]. Tessier et al. have shown that the absorption plateau of nanocrystalline $Ti_{50}Fe_{50}$ is narrower and at a lower pressure than the coarse-grained TiFe [17]. The γ phase was also absent in the nanocrystalline $Ti_{50}Fe_{50}$ hydride. The elastic stress from the amorphous phase produced during milling and chemical disorder are responsible for these effects.

9.2 Ti–Fe Alloy Synthesized by Mechanical Alloying

The investigation showed that TiFe could be synthesized by mechanical alloying of Ti and Fe, followed by annealing at a temperature higher than 700°C [18]. Figure 9.2 shows a series of XRD spectra of mechanically alloyed Ti–Fe powder mixture (53.85 wt% Ti + 46.15 wt% Fe) subjected to milling at increasing time. The originally sharp diffraction lines of Ti and Fe gradually become broader (Fig. 9.2b) and their intensity decreases with milling time. The powder mixture milled for more than 20 h is transformed completely to the amorphous phase, without formation of other phases (Fig. 9.2c). During the MA process, the crystalline size of Ti decreases with mechanical alloying time and reaches a steady value of 20 nm after 15 h of milling. This size of crystallites seems to be favorable to the formation of an amorphous phase, which develops at the Ti–Fe interfaces. Formation of the nanocrystalline alloy TiFe was achieved by annealing the amorphous material in high purity argon atmosphere at 700°C for 0.5 h (Table 9.1, Figs. 9.1a and 9.2d) [3, 18]. All diffraction peaks were assigned to those of CsCl-type structure with cell parameter a = 2.973 Å. When nickel is added to TiFe$_{1-x}$Ni$_x$, the lattice constant a increased.

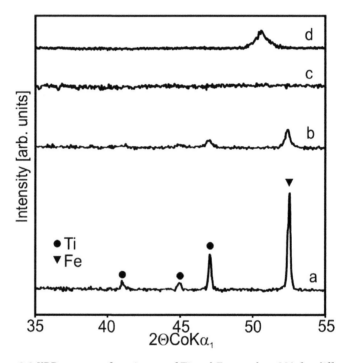

Figure 9.2 XRD spectra of a mixture of Ti and Fe powders MA for different times in argon atmosphere: (a) initial state (elemental powder mixture), (b) after MA for 2 h, (c) after MA for 20 h, and (d) heat treated at 700°C for 0.5 h.

The electrochemical pressure-composition (e.p.c.) isotherms for absorption and desorption of hydrogen were obtained from the equilibrium potential values of the electrodes, measured during intermittent charge and/or discharge cycles at constant current density, by using the Nernst equation [19]. Due to the amorphous nature of the studied alloys prior to annealing, the hydrogen absorption-desorption characteristics are not satisfactory. Compared with nanocrystalline TiFe, its storage capacity was considerably smaller (Fig. 9.3). Annealing causes transformation from the amorphous to the crystalline structure and produces grain boundaries. Anani et al. noted that grain boundaries are necessary for the migration of the hydrogen into the alloy [20]. It is worth noting, that the characteristics for microcrystalline and nanocrystalline materials are very similar with respect to hydrogen contents, but there are small differences in the plateau pressures. When the amount of Ni in $TiFe_{1-x}Ni_x$ was increased, the pressure in the plateau region continued to decrease and the hydrogen storage capacity was increased. The hydrogenation behavior of the amorphous structure was different from that of the thermodynamically stable, crystalline material. For amorphous TiFe material, the plateau totally disappears.

Table 9.2 reports the discharge capacities of the studied microcrystalline, amorphous, and nanocrystalline TiFe materials. The discharge capacity of electrodes prepared from the TiFe alloy powder by application of MA and annealing displayed a very low capacity (7.50 mA h g^{-1} at 4 mA g^{-1} discharge current), whereas the arc melted ones had none [3, 21, 22]. The reduction of the powder size and the creation of new surfaces are effective for the improvement of the hydrogen absorption rate.

Table 9.2 Discharge capacities of TiFe alloy prepared by different methods in the third cycle (current density of charging and discharging was 4 mA g^{-1})

Microstructure	Processing method	Lattice constant a [Å]	Discharge capacity [mAh g^{-1}]
Microcrystalline	Arc melting and annealing*	2.977	0.00
Amorphous	MA	—	5.32
Nanocrystalline	MA and annealing	2.973	7.50

*900°C/3 days.

Materials obtained when Ni was substituted for Fe in TiFe lead to great improvement in the activation behavior of the electrodes. It was found that the increasing nickel content in TiFe$_{1-x}$Ni$_x$ alloys leads initially to an increase in discharge capacity, giving a maximum at $x = 0.75$ [21]. In the annealed nanocrystalline TiFe$_{0.25}$Ni$_{0.75}$ powder, discharge capacity of up to 155 mA h g^{-1} (at 40 mA g^{-1} discharge current) was measured (Table 9.3). The electrodes mechanically alloyed and annealed from the elemental powders displayed the maximum capacities at around the third cycle, but, especially for $x = 0.5$ and 0.75 in TiFe$_{1-x}$Ni$_x$ alloy, degraded slightly with cycling. This may be due to the easy formation of the oxide layer (TiO$_2$) during the cycling.

Table 9.3 Structural parameters and discharge capacities for nanocrystalline TiFe$_{1-x}$Ni$_x$ materials (current density of charging and discharging was 40 mA g^{-1})

x	a [Å]	V [Å^3]	Discharge capacity on 3rd cycle [mA h g^{-1}]
0.0	2.973	26.28	0.7
0.25	2.991	26.76	55
0.5	3.001	27.03	125
0.75	3.010	27.27	155
1.0	3.018	27.49	67

Microstructure and possible local ordering in the TiNi samples was studied by TEM [23]. The sample milled for 5 h was mostly amorphous as appears from a high-resolution image (Fig. 9.3a). SAED pattern (see inset to Fig. 9.3a) contains broad rings at positions expected for TiNi with CsCl structure. There are, however additional weak, diffuse rings, most probably from TiO$_2$. It has been found that the amorphous alloy was unstable upon exposure to electron beam and underwent some crystallization. In the image acquired after 25 min exposure to the beam (Fig. 9.3b), formation of an ordered region (defined as patches with parallel lattice fringes) is seen. Accordingly, additional sharp reflections appear in the SAED pattern (inset to Fig. 9.3b). Apart from the prevailing amorphous phase, the milled sample contained a small amount of crystalline alloy with CsCl structure (Fig. 9.3c). Lack of any sharp reflections in the XRD pattern suggests that the amount of the crystalline phase is very low and/or it forms during in TEM observation.

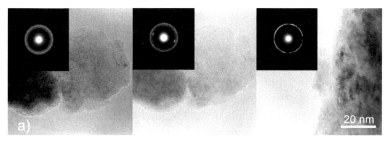

Figure 9.3 TEM micrographs and electron diffraction patterns (insets) of the milled TiNi sample: (a) typical amorphous fragment, (b) same region after 25 min exposure to electron beam, (c) crystalline grain.

The microstructure of the annealed TiNi sample is shown in Fig. 9.4. The analysis of high-resolution images (Fig. 9.4a,b) revealed the presence of well-developed crystallites with broad range of sizes from 4 up to more than 30 nm. SAED pattern obtained from large area (200 μm) (Fig. 9.4c) contains sharp rings corresponding to TiNi alloy with CsCl structure.

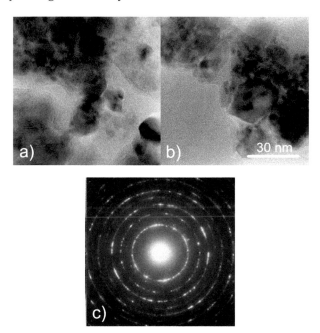

Figure 9.4 TEM micrographs (a,b) and electron diffraction patterns (c) of the annealed TiNi sample; nanocrystallites of an alloy are clearly visible in a and b.

Zaluski et al. [24] compared the hydrogenation behavior of amorphous TiFe, prepared by mechanical alloying, with that of a microcrystalline TiFe alloy. It was found that even with only a small amount of oxygen, an activation treatment was required. This might be due to the poisoning effect of oxygen which stopped the hydrogen absorption. The cleanness of the surface of microcrystalline and nanocrystalline hydrogen absorbing alloys was studied by X-ray photoelectron spectroscopy (XPS) and auger electron spectroscopy (AES) [25, 26].

On the other hand, the discharge capacity of nanocrystalline TiNi$_{0.6}$Fe$_{0.1}$Mo$_{0.1}$Cr$_{0.1}$Co$_{0.1}$ powder has not changed much during cycling (Fig. 9.5). The alloying elements Mo, Cr, and Co, substituted simultaneously for iron atoms in nanocrystalline Ti(Fe-Ni) master alloy, prevent the oxidation of this electrode material.

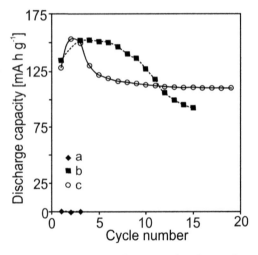

Figure 9.5 Discharge capacity as a function of cycle number of electrode prepared with nanocrystalline TiFe (a), TiFe$_{0.25}$Ni$_{0.75}$ (b), and TiNi$_{0.6}$Fe$_{0.1}$Mo$_{0.1}$Cr$_{0.1}$Co$_{0.1}$ (c); (solution, 6 M KOH; temperature, 20°C). The charge conditions were: 40 mA g^{-1}; The cut-off potential vs. Hg/HgO/6M KOH was −0.7 V.

9.3 Electronic Structure

The application of titanium alloys as electrode materials focused also on the electronic structure of TiFe and its modification by mainly Ni atoms as well as Mo, Cr, and Co impurities [27]. Earlier band structure calculations were performed by Liu and Ye [28] for TiFe and TiNi. The band structures of the TiFe-based alloys were calculated according to the TB-LMTO method in the atomic sphere approximation (ASA) [29, 30]. In this approximation, the crystal is divided into space-filling spheres, therefore with slightly overlapping spheres centered on each of the atomic sites. In the calculations reported here, the Wigner–Seitz (WS) sphere radii are such that the overlap is about 8.2–8.3%. The average WS radii (S_{av}) were scaled such that the total volumes of all spheres are equal to the equilibrium volumes of the unit cells with the lattice constants collected in Table 9.4. The input electronic configurations were taken as: core[Ar]+$3d^24s^2$ for the Ti atom, core[Ar]+$3d^64s^2$ for the Fe atom, core[Ar]+$3d^54s^1$ for the Ni atom, core[Ar]+$3d^54s^1$ for the Cr atom, core[Ar]+$3d^74s^2$ for the Co atom, and core[Kr]+$3d^54s^1$ for the Mo atom. The Perdew–Wang [31] potential with nonlocal corrections was used in the calculations and the spin–orbit interactions were taken into account in the form proposed by Min and Jang [32]. The combined correction terms [29] were included to compensate for errors due to the ASA. The Brillouin zone k-point integrations were carried out using the tetrahedron method [33] on a grid of 256 and 1331 k-points in the irreducible part (1/48 and 1/8 for $TiFe_{1-x}Ni_x$ and Ti–Fe–Ni systems doped by Mo–Cr–Co atoms, respectively) of the cubic Brillouin zone, which corresponds to 8000 k-points in throughout the Brillouin zone. The iterations were repeated until the energy eigenvalues of consecutive iteration steps were the same within an error of 0.01 mRy.

Table 9.4 Structural parameters: experimental (lattice constants a) and used in the calculations (Wigner–Seitz radii: S_A—atomic radii, where A means Ni, Fe, Mo, Cr or Co, and S_{av}—average radii). $M_x = Ni_{x/4}Mo_{x/4}Cr_{x/4}Co_{x/4}$ and $M'_x \equiv Fe_{x/4}Mo_{x/4}Cr_{x/4}Co_{x/4}$, where x = 0.5 and 0.4 for computational (S_{av}, S_{Ti} and S_A) and experimental (a, Q) values, respectively

	TiFe	$TiFe_{1-x}M_x$	$TiFe_{3/4}$ $Ni_{1/4}$	$TiFe_{1/2}$ $Ni_{1/2}$	$TiNi_{1-x}$ M'_x	$TiNi_{3/4}$ $Fe_{1/4}$	TiNi
a [Å]	2.973	2.985	2.991	3.001	3.018	3.010	3.018
S_{av} [Å]	1.4638	1.4697	1.4727	1.4776	1.4860	1.4820	1.4860
S_{Ti} [Å]	1.5013	1.4432	1.5106	1.5156	1.4607	1.5202	1.5471
S_A [Å]	1.4243	1.4954	1.4327	1.4375	1.5143	1.4418	1.4194

In the case of TiFe, the Fermi level (E_F) is located in deep valley of DOS and the value of DOS ($E = E_F$) is low, 0.354 states/(eV f.u.). As it was discussed in [25, 27] starting from the TiFe alloy and going to the TiNi one the B2 structure becomes relatively less stable, the Fermi energy shifts toward higher peaks (Fig. 9.6). The DOS ($E = E_F$) reaches the value 2.806 states/(eVf.u.). The width of the valence band is larger: 7.82 and 8.05 eV for TiFe and TiNi, respectively. The largest contribution to the DOS(E_F) is provided by the d-type electrons. The values of DOS at $E = E_F$ for $TiFe_{1-x}Ni_x$ (x = 0, ¼, ½, ¾, 1) are presented in Table 9.5. Larger number of valence electrons provided by Ni atoms move the Fermi level toward higher energies and DOS ($E = E_F$) gradually increases with increasing content of Ni atoms. The plots of DOS for $TiFe_{1-x}Ni_x$ (x = ¼, ½, ¾) are presented in Fig. 9.7. The bottom of valence band moves to lower energies and valence bands become wider. The center of gravity of the Ti d-electrons lying above E_F shifts toward lower energies close to the centers of gravity of d-electrons located on Fe and Ni atoms. Larger overlapping of bands leads to larger hybridization and charge transfer between atoms. The occupation numbers of electrons distributed between atoms are presented in Table 9.6. The analysis of these data shows that Ni atoms in $TiFe_{1-x}Ni_x$ alloys cause the outflow of electrons from Ti atoms, which is maximum for x = ¾. The strength of hybridization is shown in Fig. 9.7 on Ni projected DOS where sharp peak (mainly d electrons) for x = ¼ becomes broadened with increasing content of Ni atoms.

In order to improve the electrochemical properties of the nanocrystalline electrode materials studied so far, the ball-milling technique was applied to the TiFe-type alloys using the nickel and graphite elements as a surface modifier [34]. The $TiFe_{0.25}Ni_{0.75}$/M-type composite materials, where M = 10 wt% Ni or C, were produced by ball-milling for 1 h. Ball-milling with nickel or graphite of $TiFe_{0.25}Ni_{0.75}$-type materials is sufficient to considerably broaden the diffraction peaks of $TiFe_{0.25}Ni_{0.75}$ (not shown). Additionally, milling with graphite is responsible for a sizeable reduction of the crystallite sizes of $TiFe_{0.25}Ni_{0.75}$/C from 30 to 20 nm.

Table 9.5 Densities of states at the Fermi level [states/(eV·atom)]

Total (perf.u.)	TiFe 0.354	TiFe$_{3/4}$Ni$_{1/4}$ 1.325	TiFe$_{1/2}$Ni$_{1/2}$ 2.249	TiFe$_{1/4}$Ni$_{3/4}$ 2.468	TiNi 2.806
Ti	0.193	0.554	1.092	1.360	1.673
s	0.002	0.007	0.018	0.027	0.030
p	0.033	0.062	0.134	0.119	0.120
d	0.158	0.485	0.940	1.214	1.524
Fe	0.161	0.655	0.979	0.802	—
s	0.006	0.032	0.082	0.046	—
p	0.035	0.096	0.281	0.246	—
d	0.119	0.528	0.616	0.511	—
Ni	—	1.121	1.334	1.209	1.132
s	—	0.034	0.024	0.039	0.044
p	—	0.111	0.170	0.234	0.221
d	—	0.976	1.140	0.937	0.868

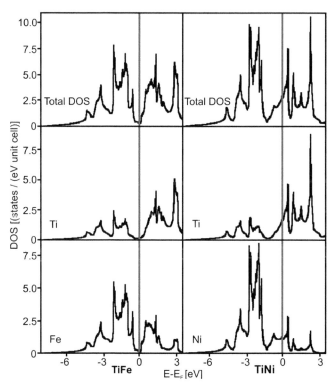

Figure 9.6 Total and local DOS plots for TiFe and TiNi systems.

Table 9.6 Numbers of states NOS [electrons/atom].

Total (per f.u.)	TiFe 12	TiFe$_{3/4}$Ni$_{1/4}$ 12.5	TiFe$_{1/2}$Ni$_{1/2}$ 13	TiFe$_{1/4}$Ni$_{3/4}$ 13.5	TiNi 14
Ti	3.8002	3.7623	3.7214	3.6837	3.7832
s	0.5714	0.5657	0.5571	0.5508	0.5759
p	0.7559	0.7397	0.7187	0.7042	0.7417
d	2.4729	2.4569	2.4456	2.4287	2.4656
Fe	8.1997	8.2115	8.2170	8.2333	—
s	0.6975	0.7157	0.7554	0.7645	—
p	0.7638	0.7457	0.7096	0.7094	—
d	6.7384	6.7501	6.7520	6.7594	—
Ni	—	10.3165	10.3390	10.3440	10.2168
s	—	0.7891	0.7873	0.8216	0.8065
p	—	0.8489	0.8494	0.8293	0.7679
d	—	8.6785	8.7023	8.6931	8.6424

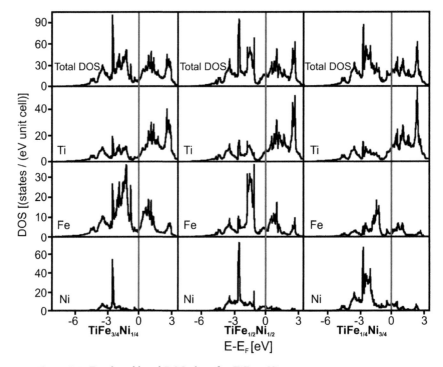

Figure 9.7 Total and local DOS plots for TiFe$_{1-x}$Ni$_x$ systems.

Figure 9.8 shows the discharge capacities as a function of the cycle number for studied nanocomposite materials. When coated with nickel, the discharge capacities of nanocrystalline TiFe$_{0.25}$Ni$_{0.75}$ powders were increased. The elemental nickel was distributed on the surface of ball milled alloy particles homogenously and role of these particles is to catalyze the dissociation of molecular hydrogen on the surface of studied alloy. Mechanical coating with nickel or graphite effectively reduced the degradation rate of the studied electrode materials. Compared to that of the uncoated powders, the degradation of the coated was suppressed. Recently, XPS investigations indicated the interaction of graphite with MgNi alloy occurred at the Mg part in the alloy [35]. Graphite inhibits the formation of new oxide layer on the surface of materials once the native oxide layer is broken during the ball-milling process.

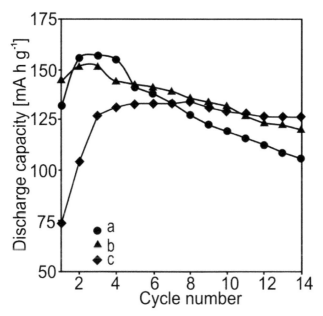

Figure 9.8 The discharge capacity as a function of cycle number for MA and annealed TiFe$_{0.25}$Ni$_{0.75}$ (a) as well as TiFe$_{0.25}$Ni$_{0.75}$/Ni (b) and TiFe$_{0.25}$Ni$_{0.75}$/C (c) composite electrodes (solution, 6M KOH; temperature, 20°C).

Materials obtained by the substitution of Ni by Fe lead to great improvement in the activation behavior of the electrodes.

It was found that the increasing nickel content in TiNi$_x$Fe$_{1-x}$ alloys leads initially to an increase in discharge capacity, giving a maximum at $x = 0.75$. Before discussing the effect of the Ni substitution by Fe on the electronic structure, we shall summarize the most important features of the experimental XPS valence band of the reference polycrystalline TiNi compound. The XPS valence band of the polycrystalline TiNi alloy is plotted in Fig. 9.9a. The photoelectron cross section of the Ti d states is significantly lower than that of the Ni 3d states.

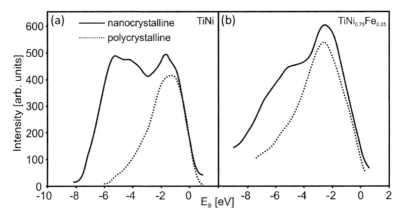

Figure 9.9 XPS valence band (Al–Kα) spectra for the nanocrystalline and polycrystalline TiNi (a) and TiNi$_{0.75}$Fe$_{0.25}$ (b) alloys. The polycrystalline samples were prepared by standard induction melting method. The nanocrystalline samples were prepared by mechanical alloying followed by annealing.

Therefore, to a first approximation, the occupied part of the valence band is dominated by the Ni 3d states. Furthermore, the Ni band peak position does not change when passing from pure Ni to TiNi. On the other hand, the intensity of TiNi XPS intensity at the Fermi level decreases strongly in the compound.

Experimental XPS valence band spectrum of polycrystalline TiNi$_{0.75}$Fe$_{0.25}$ alloy is shown in Fig. 9.9b (broken line). The corresponding structure of the valence band is significantly modified compared to that measured for the TiNi. The XPS signal at E_F is high and mostly composed of Ni-d states since the Fe-d contribution is practically negligible. In Fig. 9.9b (solid line), we show the XPS valence band of the nanocrystalline TiNi$_{0.75}$Fe$_{0.25}$ alloy. The shape of the polycrystalline sample (broken line in

Fig. 9.9b). A similar behavior was also observed in other nanocrystalline TiNi-based alloys.

The significant broadening of the valence band for the nanocrystalline TiNi-based alloys could be explained by disorder and strong deformation of the nanocrystals [25, 26]. Similar to the behavior reported above for the MA nanocrystalline Mg_2Ni-, Mg_2Cu-, and $LaNi_5$-type alloys. Furthermore, in the case of MA nanocrystalline alloys, the Ni atoms could also occupy metastable positions in the deformed grain. The above behavior can also modify the electronic structure of the valence band. As a result, the strong modifications of the electronic structure of the nanocrystalline TiNi-type alloys compared to that of polycrystalline materials could influence its hydrogenation properties.

References

1. Massalski, T. B. (1990). *Binary Alloy Phase Diagrams*. 2nd ed. Ohio: ASM International.
2. Reilly, J. J., and Wiswall, R. H. (1974). Formation and properties of iron titanium hydride. *Inorg. Chem.* **13**: 218–222.
3. Jankowska, E., and Jurczyk, M. (2002). Electrochemical behaviour of high-energy ball-milled TiFe alloy, *J. Alloys Comp.* **346**: L1–L3.
4. Schlapbach, L. (1992). *Surface Properties and Activation in Hydrogen in Intermetallic Compounds II*, vol. 67 (Schlapbach, L., ed.), Springer-Verlag, Berlin, p. 15.
5. Psoma, A., and Sattler, G. (2002). Fuel cell systems for submarines: From the first idea to serial production. *J. Power Sources* 106: 381–383.
6. Hirscher, M., ed. (2010). *Handbook of Hydrogen Storage: New Materials for Future Energy Storage*. WILEY-VCH.
7. Chiang, C.-H., Chin, Z.-H., and Perng, T. P. (2000). Hydrogenation of TiFe on high energy ball milling. *J. Alloy. Comp.* **307**: 259–265.
8. Zaluski, L., Tessier, P., Ryan, D. H., Doner, C. B., Zaluska, A., Strom-Olsen, J. O., Trudeau, M. L., and Schulz, R. (1993). Amorphous and nanocrystalline Fe-Ti prepared by ball milling, *J. Meter. Res.* **8**: 3059–3068.
9. Sun, L., Liu, H., Bradhurst, D. H., Dou, S. (1998). Formation of FeTi hydrogen storage alloys by ball-milling. *J. Mater. Sci. Lett.* **17**: 1825–1830.

10. Choong-Nyeon, P., and Jai-Young, L. (1984). The activation of FeTi for hydrogenation. *J. Less Common Metals* **96**: 177–182.
11. Trudeau, M. L., Schulz, R., Zaluski, L., Hosatte, S., Ryan, D. H., Doner, C. B., Tessier, P., Strom-Olsen, J. O., and Van Neste, A. (1992). Nanocrystalline iron-titanium alloys prepared by high-energy mechanical deformation. *Mater. Sci. Forum* **88–90**: 537–544.
12. Lee, M., Kim, J. S., Koo, Y. M., and Kulkova, S. E. (2002). The adsorption of hydrogen on B2 TiFe surfaces. *Int. J. Hydrogen Energy* **27**: 403–412.
13. Bououdina, M., Fruchart, D., Jacquet, S., Pontonnier, L., Soubeyroux, J. L. (1999). Effect of nickel alloying by using ball milling on the hydrogen absorption properties of TiFe. *Int. J. Hydrogen Energy* **24**: 885–890.
14. Zaluski, L., Zaluska, A., Tessier, P., Strom-Olsen, J. O., and Schulz, R. (1996). Hydrogen absorption by nanocrystalline and amorphous Fe-Ti with palladium catalyst produced by ball milling, *J. Mater. Sci.* **31**: 695–698.
15. Schlapbach, L., and Riesterer, T. (1983). The activation of FeTi for hydrogen absorption. *Appl. Phys. A* **32**: 169–182.
16. Zaluski, L., Zaluska, A., Tessier, P., Strom-Olsen, J. O., and Schulz, R. (1996). Investigation of structural relaxation by hydrogen absorption in ball-milled alloys, *Mater. Sci. Forum* **225–227**: 875–880.
17. Tessier, P., Zaluski, L., Zaluska, A., Strom-Olsen, J. O., and Schulz, R. (1996). Effect of compositional variations on hydrogen storage in ball-milled Fe-Ti, *Mater. Sci. Forum* **225–227**: 869–874.
18. Jurczyk, M., Jankowska, E., Nowak, M., and Wieczorek, I. (2003). Electrode characteristics of nanocrystalline TiFe-type alloys. *J. Alloys Comp.* **354**: L1–L4.
19. Kopczyk, M., Wojcik, G., Młynarek, G., Sierczynska, A., and Beltowska-Brzezinska, M. (1996). Electrochemical absorption-desorption of hydrogen on multicomponent Zr-Ti-V-Ni-Cr-Fe alloys in alkaline solution. *J. Appl. Electrochem.* **26**: 639–645.
20. Anani, A., Visintin, A., Petrov, K., and Srinivasan, S. (1994). Alloys for hydrogen storage in nickel/hydrogen and nickel/metal hydride batteries. *J. Power Sources* **47**: 261–275.
21. Jurczyk, M., Jankowska, E., Nowak, M., and Jakubowicz, J. (2002). Nanocrystalline titanium type metal hydrides prepared by mechanical alloying. *J. Alloys Comp.* **336**: 265–269.
22. Jurczyk, M., and Nowak, M. (2008). Nanomaterials for hydrogen storage synthesized by mechanical alloying, in: Ali, E., ed., *Nanostructured Materials in Electrochemistry* (Wiley), Chapter 9.

23. Makowiecka, M., Kępiński., M., and Jurczyk, M. (2008). Nanoscale hydrogen storage materials studied by TEM. *Rev. Adv. Mater. Sci.* **18**: 621–626.
24. Zaluski, L., Zaluska, A., and Ström-Olsen, J. O. (1997). Nanocrystalline metal hydrides. *J. Alloys Comp.* **253–254**: 70–79.
25. Smardz, K., Smardz, L., Jurczyk, M., and Jankowska, E. (2003). Electronic properties of nanocrystalline and polycrystalline TiFe$_{0.25}$Ni$_{0.75}$ alloys. *Physica Status Solidi* (a) **196**: 263–266.
26. Smardz, L., Jurczyk, M., Smardz, K., Nowak, M., Makowiecka, M., and Okonska, I. (2008). Electronic structure of nanocrystalline and polycrystalline hydrogen storage materials. *Rev. Energy* **33**: 201–210.
27. Szajek, A., Jurczyk, M., and Jankowska, E. (2002). The electronic and electrochemical properties of the TiFe based alloys. *J. Alloys Comp.* **348**: 285–292.
28. Liu, H. J., and Ye, Y. Y. (1998). Electronic structure and stability of Ti-based shape memory alloys by LMTO-ASA. *Solid State Commun.* **106**: 197–202.
29. Andersen, O. K. (1975). Linear methods in band theory. *Phys. Rev. B* **12**: 3060–3083.
30. Szajek, A., Jurczyk, M., Smardz, L., Okonska, I., and Jankowska, E. (2007). Electrochemical and electronic properties of nanocrystalline Mg-based hydrogen storage materials. *J. Alloys Comp.* **436**: 345–350.
31. Perdew, J. P., Chevary, J. A., Vosko, S. H., Jackson, K. A., Pederson, M. R., Singh, D. J., and Fiolhais, C. (12992). Atoms, molecules, solids, and surfaces: Applications of the generalized gradient approximation for exchange and correlation. *Phys. Rev. B* **46**: 6671–6687.
32. Ming, B. I., and Jang, Y. R. (1991). The effect of the spin-orbit interaction on the electronic structure of magnetic materials. *J. Phys.: Condens. Matter.* **3**: 5131–5561.
33. Blöchl, P., Jepsen, O., and Andersen, O. K. (12994). Improved tetrahedron method for Brillouin-zone integrations. *Phys. Rev. B* **49**: 16223–16233.
34. Jurczyk, M., Smardz, L., Makowiecka, M., Jankowska, E., and Smardz, K. (2004). The synthesis and properties of nanocrystalline electrode materials by mechanical alloying. *J. Phys. Chem. Sol.* **65**: 545–548.
35. Smardz, L., Jurczyk, M., Smardz, K., Nowak, M., Makowiecka, M., and Okonska, I. (2008). Electronic structure of nanocrystalline and polycrystalline hydrogen storage materials. *Renew. Energy* **33**: 201–210.

Chapter 10

TiNi-Based Hydrogen Storage Alloys and Compounds

Mateusz Balcerzak

*Institute of Materials Science and Engineering,
Poznan University of Technology, Poznan, Poland*

mateusz.balcerzak@put.poznan.pl

10.1 Phase Diagram and Structure

A Ti–Ni binary phase system contains three intermetallic stable phases—Ti_2Ni, $TiNi$, and $TiNi_3$—and several metastable intermetallic phases [1] (Fig. 10.1). The Ti–Ni system belongs to a typical Brewer hypo-hyper-d-electronic combination of transient metals. It characterizes the transfer of the paired d-electrons from filled semi-d-orbitals to an empty d-subshell of Ti (partial d-band transfer) and thereby a strong intermetallic bonding [2].

Hydrogen storage properties of Ti–Ni materials have been investigated by many researchers during last decades. Ti–Ni alloys composed of Ti_2Ni and $TiNi$ phases were among the first alloys studied as negative electrodes for possible use in nickel-hydride batteries [3]. It was due to the low cost of the system. Moreover, Ti_2Ni and $TiNi$ alloys actively absorb and desorb hydrogen at room temperature [4]. On the other hand, $TiNi_3$

Handbook of Nanomaterials for Hydrogen Storage
Edited by Mieczyslaw Jurczyk
Copyright © 2018 Pan Stanford Publishing Pte. Ltd.
ISBN 978-981-4745-66-6 (Hardcover), 978-1-315-36444-5 (eBook)
www.panstanford.com

has good catalytic activity for hydrogen adsorption/desorption [4, 5]. Electrocatalytic activity of Ti–Ni-type materials depends on the elemental composition and decrease in the order $TiNi_2 > TiNi_3 > TiNi_4 > Ti_2Ni > TiNi > Ti_3Ni > TiNi_{0.7}$ [2].

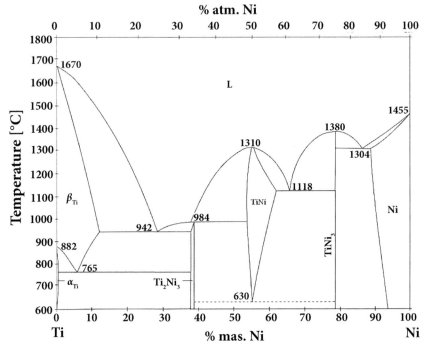

Figure 10.1 Phase diagram of Ti–Ni alloy by Massalski et al. [8].

One of Ti–Ni alloys is Ti_2Ni alloy. The melting point of Ti_2Ni is 984°C [6, 7]. Ti_2Ni is non-magnetic material, and possesses metallic anisotropic nature and behave like a ductile material [7]. This phase has face-centered cubic (fcc) structure. The space group is Fd-3m (number 227) [7]. Lattice constant calculated equals 11.3193(2) Å [3]. The titanium atoms occupy the 16c and 48f lattice site positions and the nickel atoms occupy the 32e position. The structure possesses two internal parameters, u and v, which determine the non-equivalent position of Ni (u,u,u) and Ti_2 (v, 1/4, 1/4). The number of atoms per unit cell is 96 [7].

Ti_2Ni alloy is a promising anode material for Ni-MH batteries and a material for gaseous hydrogen storage due to its relatively high hydriding-dehydriding kinetics, ability to remain stable in an electrolyte and reasonable specific capacity [6]. Moreover, Ti$_2$Ni alloy can be used in the form of hydrides acting as moderators for space nuclear auxiliary power systems or for hydrogen purification [9]. Ti$_2$Ni offers excellent fast discharge performance because of high diffusion of H atoms within its crystal lattice [10]. The activation energy for hydrogen diffusion in the Ti$_2$Ni-based electrode is calculated to be 10.7 kJ/mol [11]. This alloy reacts with hydrogen creating four hydride phases: Ti$_2$NiH$_{0.5}$, Ti$_2$NiH, Ti$_2$NiH$_2$, and Ti$_2$NiH$_{2.5}$ [9, 12]. It is important to note that Buchner et al. first reported that metal hydride Ti$_2$NiH$_{2.9}$ (1.86 mass% H), which corresponds to an electrochemical capacity of 500 mAh/g could be formed after gaseous hydrogen absorption [13, 14]. Such a high capacity is due to high number of interstitial sites for hydrogen storage [15]. Unfortunately, Ti$_2$Ni hydride is to stable to desorb hydrogen at ambient temperature and under atmospheric pressure [16].

As already mentioned in this chapter, another Ti–Ni alloy which is able to absorb hydrogen is TiNi alloy. This material is characterized by low specific weight, high hydrogen capacity per unit weight, and good oxidation resistance, which is desirable in hydrogen storage materials [7]. Unfortunately, TiNi charge/discharge kinetics is slow [16]. The enthalpy of hydrogen solution per H atom was determined to be 183 ± 40 meV and the activation energy of hydrogen diffusion −480 ± 50 meV [17].

TiNi crystallizes in the cubic CsCl-type structure [18]. The cell parameter of this phase equals a = 3.019 Å [18, 19]. Enthalpy of formation per atom for TiNi alloy −0.373 eV/atom [20]. TiNi during hydrogenation reacts with hydrogen to form a hydride with composition close to TiNiH. This hydride is crystallizing in a I4/mmm space group [1].

Moreover, TiNi alloy has many other interesting properties such as shape memory effect, high melting temperature, and increased hardness [7]. The shape memory properties are based on a reversible martensitic transformation [21]. For binary TiNi, the martensitic phase starts to nucleate on cooling at 47°C while heating the austenitic phase starts to form at 67°C [22]. As shape memory alloy, TiNi has lots of mechanical and biomedical application—actuators in micro-electromechanical systems (MEMS), piping, actuators, stents [7]. Moreover, it finds wide application in aerospace and automotive industries due to the high corrosion resistance [7, 21]. TiNi alloy found also application as a MOS capacitive sensor and as highly reflecting mirrors, supermirrors, monochromators, and polarizers in the field of soft X-ray and neutron optics. TiNi is also known as superelastic material [7].

As was written above, both, TiNi and Ti_2Ni alloys possess some drawbacks with regard to their use as a hydrogen storage material or electrode material in $Ni-MH_x$ batteries. Therefore, there are methods which can be used to improve properties of these alloys, such as oxygen addition, chemical substitution, destructive hydrogenation recombination technique, hydrogen activation, or microstructure modification [1].

10.2 Electrochemical and Gaseous Hydrogen Sorption Measurements

Materials with potential use as a Ni-MH rechargeable batteries are electrochemically tested to check they properties: discharge capacity, cycle stability and others. Most often the result of the research is obtaining the cycle-life curves (discharge capacity vs. cycles of charging/discharging).

The efficiency of the hydrogen storage alloy electrode is determined by both the kinetic of the process occurring at the metal/electrolyte interface and the rate of hydrogen atom diffusion within the bulk of the alloy [4]. The discharge process of Ti–Ni alloys electrodes which is exposed to 6 M KOH solution is composed of three steps:

(a) diffusion of atomic hydrogen from inside the bulk alloy to the surface of the electrode

(b) adsorption of atomic hydrogen on the surface of the electrode

(c) oxidation of atomic hydrogen which generates electric current [23]

Discharge capacity of TiNi alloy is influenced by the maximum discharge capacity of the alloy and the discharge kinetics which is determined by the electrochemical polarization and concentration polarization. This depends on the diffusion ability of hydrogen and the rate of phase transformation. Cycle life at low cycling current density is mainly determined by the decline in the maximum discharge capacity, but when cycling current density are high, the cycle life is controlled by the decline in the discharge kinetics. Increase in the discharge current density and cycling current density decrease the discharge capacity but improve the cycle life behavior [24].

10.2.1 Cycle-Life Curves

Cycle-life curves can be split in two parts (see Fig. 10.2). A first one is named activation period, where the discharge capacity increases until the maximum capacity is reached. Moreover, the intermetallic alloy is said activated when also the experimental value of its half-discharge potential became constant and the polarization of the electrode of this alloy became minimum and constant, independently of the cycling [15]. The activation is mostly controlled by phase transformation [24]. The second one is named degradation period, where the discharge capacity decreases [25].

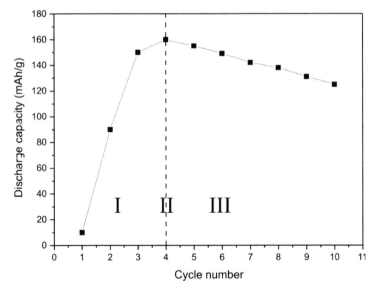

Figure 10.2 Schematic pattern of discharge capacity as a function of cycle number: (I) activation period, (II) maximum discharge capacity, (III) degradation period.

10.2.1.1 Electrochemical activation process in Ti–Ni alloys

The number of cycles needed to reach the activated state of alloy is dependent of each material and can range from a few to a several cycles of charging-discharging [26]. For example, TiNi alloys chemically modified by B need about 40 cycles to reach maximum value of discharge capacity [27].

It is also important to note that many researchers carry out the activation process before electrochemical test in order to obtain better electrochemical properties of studied materials. For example Ti–Ni-based alloy can be activated by charging and discharging at 25 mA/g to the end point [28, 29]. Other group dipped active material in a 10% HF solution for 10 min [30].

As an electrolyte for this kind of electrochemical test a KOH is used. This electrolyte plays a major role on the cycle-life behavior of the electrodes. Moreover, it affects activation, which shortens with increasing the KOH concentration [25].

10.2.1.2 Activation in gaseous hydrogenation measurements

Activation method in gaseous hydrogenation measurements is based on initial hydrogenation of studied material under high pressures and/or temperatures. Activation process is intended to obtain maximum H-capacity and faster hydrogenation/dehydrogenation kinetics in materials. Initial H-penetration depends on barriers and surface structure. Hydrogenation causes fracturing of alloy material, which results in formation of a large number of small pieces, increasing fresh reaction surface area [1].

Activation of Ti–Ni systems causes reduction in particle size due to hydrogen decrepitation, which is self-pulverization into smaller particles of material located in hydrogen. This phenomenon results from a combination of volume expansion due to the hydride formation and a brittle nature of hydrides. As a result of this process particle size distribution have become more uniform. It is worth noting, titanium alloys have tendency to disintegrate into fine particles. However, pulverization during hydrogenation of the Ti_2Ni alloy is easier than for TiNi alloy. As recent research has shown, activation process can increase the hydrogen capacity by 42% and by even 400% in case of TiNi and Ti_2Ni respectively [1].

The activation process also can affect kinetics of hydrogen sorption. Time of hydrogen sorption can be reduced dramatically after two cycles of activation. Ti–Ni-type alloys absorbed hydrogen without any incubation time. For TiNi alloy, not-activated powder absorbed 95% of maximum hydrogen volume in 163 min and after two activation cycles just in 7 min. This situation is similar for Ti_2Ni alloy for which the time interval decreases from 122 min to only 5 min [1].

10.2.1.3 Electrochemical and gaseous capacity degradation

Normally, Ti–Ni-based materials do not retain their maximum discharge capacity during cycling. It is known that oxidation and disintegration are prominent factors that are associated with the capacity loss during charge/discharge cycles [12, 31]. Oxidation decreases the amount of hydrogen-storage elements, behavior which is called thermodynamic degradation [24, 32]. It is important to note that 6 M KOH is a very strong oxidation electrolyte [33]. O atoms occupy the irregular octahedral interstitial sites of Ti_2Ni alloy (crystallographic position 8a). It results in the formation of Ti_4Ni_2O alloy which has the same crystal structure as Ti_2Ni alloy. Formation of Ti_4Ni_2O does not affect the expansion of Ti_2Ni lattice [11, 34]. Oxygen content of the $Ti_4Ni_2O_x$ compound varies from $x = 0$ to $x = 1$ [34]. However, Zhao et al published that the product of corrosion possesses a higher content of O than Ti_4Ni_2O resulting in increase of lattice parameter [11]. The oxide film acts as a barrier to the hydrogen diffusion and limits the hydrogen atoms transfer from the bulk to the surface [4]. The severe oxidation on the alloy surface reduced the surface activity, leading to an increase of the energy barrier for the reaction on the alloy surface [14]. Studies revealed also that TiO_2 layer of the alloy after cycling is thicker and more stable than that of as-prepared alloy. It decreases the charge efficiency and limit the hydrogen penetration though the surface [35]. Some researchers' group published that doping the passive TiO_2 film with small amount of Pt or Pd can drastically increase the catalytic properties of the film, due to the formation of impurity band conductivity in the TiO$_2$ [32]. It is also important to note that the natural oxide layer with thickness varying from 4 to 15 nm and mainly composed of TiO$_2$ and TiO is formed on the air-exposed TiNi compound at room temperature [25].

The way of electron transfer at film-covered electrodes depends on the thickness of the oxide film. When the film is sufficiently thin, electrons exchange with the underlying metal itself. The oxide film acts as a barrier for electron transfer and the current decreases strongly with the film thickness [32].

Oxidation resistance can be increased greatly by low-temperature (25°C) micro-encapsulation process [12, 36]. Nickel and copper on the surface of the Ti–Ni electrode increase the electroconductivity of the electrode and therefore increase the charge-discharge efficiency and discharge capacity. However, the discharge capacity of Ti$_2$Ni microencapsulated with nickel was higher than Ti$_2$Ni microencapsulated with copper. It is probably due to the differences between nickel and copper resistance between the alloy particle and layer. The best electrochemical properties was obtained when nickel content reached 7 wt%. Nickel on the surface enhances the electroconductivity of electrode, decreases the internal resistance of electrode and prevent oxygen from diffusing to the internal part of alloy decreasing the oxidation rate of the alloy [37].

Also formation and accumulation of Ti$_2$NiH$_{0.5}$ during work time of electrode can be directly linked with capacity loss [12]. Moreover, this hydride is easily oxidized to Ti$_4$Ni$_2$O, leading to a capacity loss during cycling [5]. Capacity loss during early cycles is dominated by formation and accumulation of Ti$_2$NiH$_{0.5}$, while oxidation and disintegration become the determining factor during later cycles [12].

Cyclic gaseous hydrogen sorption and desorption can also have negative consequences on hydrogen storage capacity, which is the damage caused by gaseous impurities. There are three types of damage: poisoning, retardation, and reaction in which presence depends on an alloy impurity combination. Another possible process which could cause negative impact on the hydrogenation properties is disproportionation. It can occur even if very pure H$_2$ is used. It is connected with tendency of metastable alloys and compounds to break up metallurgically to form stable, hardly reversed hydrides [1].

10.2.2 Electrodes and Electrochemical Measurement Conditions

Before electrochemical tests, Ti–Ni-base active material is used to make electrode for measurements. The active material is usually mixed with different type of materials (also in different ratios). Active powder can be mixed in a weight ratio with

(a) poly(vinyl alcohol) solution at a ratio of 10:1 [9, 12, 23, 33, 38],
(b) 5 mass% of polyvinyl alcohol together with metallic binder (Co, Ni and Cu) [31],
(c) nickel powder at a ratio of 1:4 and 10% poly(vinyl alcohol) solution [6],
(d) nickel at a ratio 1:4 [4, 11, 14, 38, 39],
(e) nickel at a ratio 2:7 [5, 40],
(f) nickel at a ratio 9:1 [3],
(g) carbonyl nickel at a ratio 1:5 [41],
(h) tetracarbonyl nickel at a ratio 9:1 [18],
(i) copper at a ratio 1:4 [42],
(j) copper at a ratio 1:1 and 7:3 [3],
(k) teflonized carbon black at a ratio 10:7 [16, 17, 27, 42],
(l) carbon black and polytetrafluoethylene (PTFE) at a ration 90:5:5 [21, 22].

Each of additives individually influences the electrochemical properties of Ti–Ni materials. For example, the use of copper binder leads to increase in cyclic stability with a slight decrease of the initial capacity [3]. On the other hand, the use of polyvinyl alcohol with metallic binder decreases the discharge capacity. It is due to the decrease in the reversibility of charge-discharge process and slow activation of the specimen [31].

As was written above, different research groups use different types of activation methods and additives to active material. Also the measurement conditions can be different for each research group. The values of discharge capacity also depend on charging and discharging current. Ti–Ni-based materials were measured under very wide group of discharge current density values: 20 mA/g [12, 38], 25 mA/g [29], 40 mA/g [15, 19, 43–45], 60 mA/g [11, 38, 39], 100 mA/g [13], 120 mA/g [11], 125 mA/g [29],

180 mAh/g [11], 240 mA/g [11], 300 mA/g [11], 900 mA/g [4], 1500 mA/g [4]. It is known that with increase of discharge current density the discharge capacity decreases. It is due to the depletion of atomic hydrogen from the surface of the particles at high discharge rates which causes the potential to drop before all the hydrogen contained in the particles has been reacted [4].

Moreover, the cutoff potential can be, for example, 0.2 V [12], 0.6 V [4, 11, 29, 37], 0.7 V [13, 34, 38, 46, 47], 0.74 V [49] versus Hg/HgO reference electrode. The discharge capacity increases with decrease of the cutoff potential.

10.2.3 The Influence of Temperature on Hydrogen Sorption/Desorption Properties

The hydrogenation process strongly depends on the temperature. Increase of temperature affects both the sorption and desorption of hydrogen. First, it raises the energy of the adsorbed hydrogen which affects the reduction of adsorption stability [23]. Moreover, it also affects the decay of the discharge capacity which is due to the corrosion of active material in the alkaline electrolyte [10, 11]. Importantly, adsorption of hydrogen must be exothermic if it is to proceed spontaneously [11, 23]. For these reasons, higher temperatures are not favorable to the adsorption process. On the other hand, the increasing in temperature is favorable for the dehydrogenation process. Increased temperature leads to faster supply of hydrogen from inside the hydrogen storage alloy, which results in increased gradient of thermodynamic chemical potential of hydrogen [23, 26, 29]. Moreover, charge retention increases and self-discharge rate decreases with an increase in temperature [28]. As an example, the Ti–Zr–Ni alloy revealed the best properties in 40°C [14]. In the case of Ti_3Ni_2 alloy electrode, maximum discharge capacity increases as the environment temperature increases. Unfortunately, it decays very fast, especially at higher temperatures: after a few cycles [28]. The maximum discharge capacity of $Ti_{1.4}V_{0.6}Ni$ electrode increased from 271 mAh/g at 30°C to 404 mAh/g at 70°C [41].

The kinetic properties of absorption and desorption of hydrogen can be also affected by temperature increase due to the subsequent growth of the active surface of the electrode. Unfortunately, it also accelerates the oxidation process through the formation of an oxide phase on the electrode surface. Finally it affects degradation of kinetic properties and reduces the active surface [26].

Another effect of rise of temperature on hydrogenation properties is the reduction of activation cycle number of Ti–Ni alloys [28].

10.3 Arc Melted Alloys

Ti–Ni-based alloys can be produced by many methods. One of them used first is arc melting. Most of the works on Ti_2Ni alloys produced by this method under argon protection were published by Luan's group. Obtained ingots were crushed and ground into a powder of below 100 mesh [6, 12, 23]. Unmodified Ti_2Ni alloy was characterized by 160 mAh/g. However, the capacity was lost during first three cycles to reach almost zero mAh/g [12]. Since then, many approaches such as surface modification and elemental substitution have been successfully used to improve the specific capacity and cycle life of Ti–Ni-based alloys.

10.3.1 Chemical Modification of Arc Melted Alloys

In the past, scientists checked the influence of the substitution of Ni by Co (in different amounts) in Ti_2Ni alloy on its electrochemical properties. It is important to note that Co and Ni have similar electron configuration, atomic radius, and atomic volume. Results show that cobalt addition does not change the lattice parameters of Ti_2Ni alloy. For this reason, it can be assumed that Co atoms exist at the same positions which were originally occupied by nickel atoms in the structure. Electrochemical tests have shown the that capacity retention rate of electrodes with cobalt is much higher than that of unmodified Ti_2Ni alloy. The reason for improved electrochemical properties can be the fact that cobalt addition enlarges the lattice cell and reduces the lattice expansion and contraction during hydrogenation and dehydrogenation. It results in slowed down powder disintegration process. Moreover, cobalt is effective in increasing the oxidation resistance of Ti–Ni alloy during cycling. Unfortunately XRD studies revealed formation and accumulation of irreversible Ti$_2$(NiCo)H$_{0.5}$ phase which results in decrease in discharge capacity during cycling [38].

Electrochemical properties of Ti–Ni alloys were also improved by partial substitution of Ni by Al [33]. Authors used Al because is lighter, cheaper than most metals and has good passivity under some circumstances [11]. Measurements revealed that specific capacity decrease and cycle life increase with aluminum addition. It can be explained by two factors. First, a part of Al captures some of Ti atoms and leads to the formation of Ti$_2$Al phase. This phase is not able to store hydrogen and therefore results in a decrease in the capacity per unit weight of alloy. The specific capacity can be also decreased by slowed down electrode kinetics which is caused by passivation layer (increased by substitution of Ni by Al). On the other hand, passivation layer covered the material from the electrolyte, reducing the oxidation of the rest of alloy and therefore increasing the cycle life of electrode [33].

Since potassium is known as one of the most active elements with a negative electrode potential, this element was also used by Luan et al. for partial substitution of Ni in Ti$_2$Ni alloy [50]. Because potassium is difficult in use, authors used as a starting material potassium borohydride. Authors observed that the specific capacity of the electrode increases and the cycle life of the electrode is extended with the increase of K-B addition. This improvement is connected with the fact that boron addition causes formation of defects which act as active sites, finally increasing the electrochemical activity of Ti$_2$Ni alloy. Moreover, boron does not have any effect on the specific capacity decay rate of the electrodes. On the other hand, potassium addition results in an increase in cycle life because potassium is easy to oxidize. Therefore, authors concluded that potassium atoms act as a micro-anode, while Ti$_2$Ni alloy acts as a micro-cathode. In such an arrangement, the micro-cathode is protected from oxidation by the micro-anode. This results in extended cycle life of the electrode. However, it is important to note that the potassium content in alloy reduces during cycles and therefore may lose its protection effect for further cycling [50].

10.3.2 Composites Containing Arc Melted Alloys

A mixture of equimolar amounts of TiNi and Ti$_2$Ni was prepared in the past. This mixture was characterized by discharge capacity at a level of 320 mAh/g [34]. This good properties are due to the catalytic effect of TiNi alloy [34].

Moreover, in the past, Luan's group produced Mg$_2$Ni-x wt% Ti$_2$Ni composites using "particle inlaying" method. SEM pictures revealed that the composite is composed of small Ti$_2$Ni particles which are distributed homogeneously on Mg$_2$Ni particles. Both types of particles were attached to each other by strong metallurgical joint which results from heat treatment. This kind of joint between particles was also proved by XRD measurements. Both particles are composed of monophases. However, at the interface of Mg$_2$Ni and Ti$_2$Ni, two newly formed TiNi and Ti-Mg phases occurred [6]. Authors observed that with increasing content of Ti$_2$Ni alloy in the composite, the discharge capacity increases up to 170 mAh/g for 40% of Ti$_2$Ni alloy. The observed situation can be attributed to at least two factors. The first one is related to Ti$_2$Ni alloy, which acts as a hydrogen storage material itself. The second one is correlated to the redox reaction of hydrogen which proceeds much more easily at the surface of Ti$_2$Ni alloy (since that alloy has a relatively low active energy of reaction with hydrogen and is relatively stable in alkaline solution compared to Mg$_2$Ni alloy). It means that Ti$_2$Ni particles provide active sites for the redox reaction of hydrogen, improving electrochemical properties of Mg$_2$Ni alloy. Moreover, Ti$_2$Ni alloy inlaid on the Mg$_2$Ni surface protects from oxidizing during discharging and provides the pathway for the diffusion of hydrogen.

10.4 Mechanically Alloyed Alloys

As already mentioned in the chapter, Ti–Ni alloys were mostly arc melted in the beginning. Since then, researchers have produced Ti–Ni-based materials by many methods. One of them is mechanical alloying. The electrochemical or gaseous hydrogenation properties depend on many factors in this method. For example, total hydrogen uptake and hydrogen onset temperature depend on the ball-milling time and the ball-to-powder mass ratio in some cases [51].

It is important to note that mechanically alloyed Ti–Ni alloys are characterized by better electrochemical properties than arc melted ones. This is due to many factors, one of which is the microstructure. Particle size is one of the most important factors which affects discharge properties of alloys. A smaller particle size (produced by mechanical alloying method), characterized by larger active surface area, is beneficial for discharge properties. Moreover, many micro-cracks forms in the alloy powders, during charge and discharge cycles, leading to the pulverization of the powders and finally generates fresh active surface area for hydrogen absorption and desorption [11].

Mechanical milling of a Ti–Ni-based mixture of elemental powder most frequently leads to obtaining of the amorphous phase than the extremely fine nanocrystalline phase [16]. Obtaining of nanocrystalline Ti_2Ni alloy can be facilitated by use of hydrogen as catalyst at elevated temperatures [51]. Obtaining of amorphous phase is usually confirmed through XRD measurements. When the material becomes amorphous, the XRD pattern becomes smooth and element peaks disappear. Figure 10.3 shows a schematic XRD patterns of (a) a mixture of Ti and Ni elements powders before milling and (b) the material after 8 h of milling. Wavy shape of patterns between 20 and 35° is assigned to amorphous holder material. TiNi milled for 40 h possesses a broad peak center at 42°, showing a sign of amorphization of TiNi phase [20]. It is important to note that differentiation between a "truly" amorphous material and an extremely fine-grained material in which very small crystals are embedded in an amorphous matrix has not been easy on the basis of the diffraction data. Only a supplementary investigation by neutron diffraction unambiguously confirmed that the phase produced by MA are truly amorphous [18]. The amorphous phase was also confirmed elsewhere by TEM microscopy technique [5]. Extended milling can reduce the grain size of TiNi by repeatedly fracturing until the amorphous phase is finally developed [20]. The time of amorphization depends on many factors and in case of Ti_2Ni alloy can last 8 h [19], 10 h [39], 60 h [26, 40, 51], and 100 h [5]. However, sometimes even quite long milling time does not lead to obtaining fully amorphous material. For example, milling of Ti–Ni–Mg alloy for 36 h resulted in significant reduction of Mg peaks intensity, but titanium and nickel peaks are still visible [48]. TiNi alloy can transform completely to the amorphous phase even in 5 h [18]. Differences in milling time may be related to many factors: type of mill, ball-to-powder weight ratio, rotation speed, type of milled powders, and volume of reaction chamber.

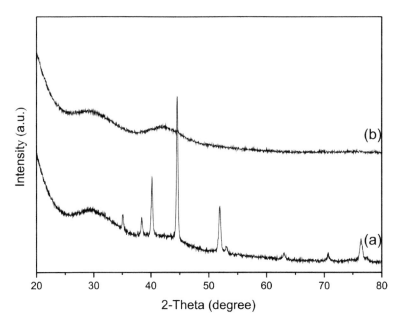

Figure 10.3 XRD patterns of: mixture of Ti and Ni elements powders (a), amorphous Ti–Ni material after 8 h of mechanical alloying.

Heat treatment is mostly done to obtain crystalline phase from amorphous phase. The main effect of the annealing on the hydriding/dehydriding process is the release of microstresses and grain growth, resulting in an increase of the well-established diffusion path for the hydrogen atoms along the numerous grain boundaries [17]. Moreover, heat-treated materials are characterized by the activation phenomenon which is attributed to powder pulverization, which increases the surface area of electrode [39]. The crystallization behavior of amorphous materials shows a multi-stage crystallization mode consisting of nucleation and growth. From AFM measurements which were done after

MA and heat treatment obtained was average size of some Ti–Ni phase crystallites—28 nm [18]. According to other AFM studies the average crystalline size of the nanocrystalline TiNi powders was of the order of 25 nm [52]. An example AFM image of Ti–Ni alloy is shown in Fig. 10.4.

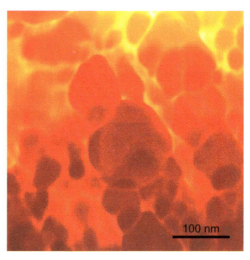

Figure 10.4   AFM picture of Ti–Ni alloy.

Most often the synthesis procedure containing MA process is composed of MA (first) with or without heat treatment. However, some researchers developed completely the opposite method of Ti$_2$Ni alloy production. In the first step, Zhao et al. mixed and cold-pressed powders of TiH$_2$ and Ni. Moreover, it was demonstrated elsewhere that the use of titanium hydride instead of pure titanium powder can shorten the amorphization time [40]. In the next step, they sintered pellet which was further ball milled for a different duration—up to 10 h [38, 39]. Alloy which was not milled was characterized by the highest capacity—278 mAh/g. However, the discharge capacity was drastically decreased during cycling. Ten hours of milling led to an amorphous phase. This alloy, which contains metastable phases, can effectively restrain capacity loss and/or the formation of irreversible phases. Mechanical milling is beneficial in improving the charge-transfer reaction (related to specific surface area and active atoms of the alloy surface) at the electrode/electrolyte interface. Moreover, ball milling leads to an increase in vacant sites which are beneficial for hydrogen movement or transportation, finally increasing hydrogen diffusion. On the other hand, longer ball milling leads to the decrease of maximum discharge capacity [38].

Summarizing, as many studies have shown, the nanocrystalline phase is more beneficial for hydrogen storage capacity, while the amorphous phase mainly contributes to kinetics improvement of hydrogen absorption-desorption and also enhancement of corrosion resistance (due to lower corrosion current density and positive corrosion potential) [5, 35, 39]. The amorphous phase of Ti_2Ni alloy helps to stabilize cycling performance [5, 26, 51]. Moreover, amorphous TiNi alloy were found to be very easily activated as well [17]. However, amorphous TiNi facilitates diffusion of hydrogen; it also leaves fewer hydrogen storage sites, resulting in general capacity decrease [20].

Electrochemical properties of mechanically alloyed Ti–Ni alloys can be improved by chemical modification. Zr element is one of the most often used elements for this purpose [14, 29]. Partial replacement of Ti by Zr resulted in the formation of new minor phase—Laves phase (λ_1)—with a $MgZn_2$-type structure [14]. This alloy can be fully activated after one cycle. Chemically modified Ti–Zr–Ni alloy possesses higher discharge capacity than unmodified Ti–Ni alloy. However, it was revealed that the formation of the Laves phase would decrease the discharge capacity. This can be due to the fact that Zr acts as a catalyst to the Ti oxidation reaction [51]. Addition of Zr decreased the crystalline temperature of the amorphous alloy, indicating that the stability of the amorphous Ti–Zr–Ni alloy was lower than that of the amorphous Ti_2Ni alloy [14]. Moreover, Zr substitution decreases the stability of the amorphous phase of Ti_2Ni. Zr addition causes flaky and inhomogeneous in size distribution in contradiction to homogenous particles of Ti_2Ni alloy particles. After addition of Zr, massive agglomeration can be observed [51].

Other research revealed that some replacement of Ti by Zr in TiNi cubic structure increases of lattice constant leading to cell-volume expansion of TiNi [51].

Chemical modification by Pd affected cyclic stability of Ti_2Ni and TiNi alloys, which can be explained by the decrease of the thickness of TiO_2 oxide film, which is the barrier for electron transfer [3, 17, 46]. Pd atoms may produce localized states in the band gap of the Ti$_2$O film. Moreover, Pd is characterized by high electrochemical catalytic activity in charge-discharge cycling. Pd-containing electrodes show high discharge capacity on cycling and high-rate dischargeability after 100 cycles [32]. It is important to note that Pd is used in small quantities due to its price.

Moreover, substitution of Ni by Pd to form TiNi$_{1-x}$Pd$_x$ compounds stabilizes the orthorhombic B19 phase as compared to the monoclinic B19 for $x > 0.15$. Moreover, owing to the higher atomic radius of Pd as compared to Ni, the cell volume of B2 and B19 structures increases linearly with Pd content. However, in spite of the larger cell volume for the host structure of intermetallics, the hydrogenation capacity and the stability of their hydrides decrease with Pd content. This behavior is attributed to electronic effect induced by Pd substitution [22].

Addition of Ag to Ti$_2$Ni phase did not change the structure of alloy very much. As XRD analysis has shown, besides the main phase, the minor TiNi phase also occurs [45]. A similar situation was also observed in case of the addition of Ag to TiNi phase. In addition to the main phase, two minor phases were detected: Ti$_2$Ni and TiNi$_3$. Part of silver in alloy could react with Ti and Ni creating Ti–Ni-based phases [44]. Silver addition does not affect the average crystallite size of alloy. Ag addition improved the corrosion resistance of materials in 6 M KOH solution. It may be caused by suppressed pulverization of electrode due to the anticorrosion effect. The effect of chemical modification by Ag is different in case of Ti$_2$Ni alloy than in case of TiNi alloy. The Ti$_2$Ni alloy with Ag addition is characterized by increased maximum discharge capacity and decreased cycle stability [6]. The situation is opposite for the TiNi-based material, which is characterized by the improved cycle stability [45]. The discharge capacities as a function of cycle number of modified Ti–Ni alloys is are presented in Fig. 10.5.

Moreover, Ti–Ni alloys were improved by chemical modification by Ni–P coating and substitution by Co, Cr, Al, Fe, Mo, Mn, Sn, Mg, Nb, V, Pd, and MoCo$_3$ [10, 29].

The cycle life of the Ti$_2$Ni electrode was greatly increased with the addition of Al, Co, Fe, Mg, and Nb [46]. MoCo$_3$ increases the initial performance of Ti$_2$Ni alloy, but the capacity decreases quickly on cycling [32]. Also, partial substitution of Ti by V in Ti$_2$Ni can improve electrochemical properties due to high affinity for hydride formation of V and easy decomposition of V hydrides. The energy barrier for H desorption from a free V surface is only 0.04 eV. V was regarded as an "atomic hydrogen pump" which facilitated atomic hydrogen transportation. Ti–V–Ni alloys can be activated at two charging/discharging cycles [41].

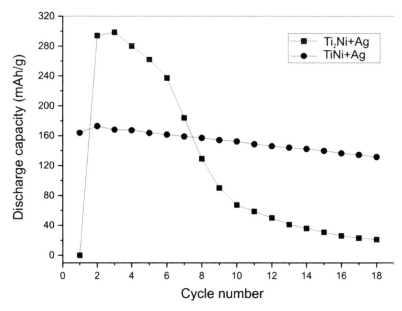

Figure 10.5 Discharge capacities as a function of cycle number of electrode prepared with Ti–Ni alloys chemically modified by Ag.

The addition of Co is effective in improvement of activation of Ti$_3$Ni$_2$ electrode alloy at all temperatures. Addition of Al decreases the charge retention of Ti$_3$Ni$_2$ alloy slightly at room temperature and greatly at high temperatures. Cr, Co, and especially Al improve significantly the cycle stability of Ti$_3$Ni$_2$ electrode alloy at room temperature [29].

Substitution of nickel by Fe, Mo, Mn, Mg, Co, and Cr in the TiNi material leads to an increase in discharge capacity in comparison to TiNi alloy [17, 18]. Moreover, substitution of Ni for Fe in TiNi$_{1-x}$Fe$_x$ leads to great improvement in the activation behavior of the electrode [18]. Substituting Nb or Mn for Ni enhances the cycle life of TiNi electrodes as well. Addition of

Sn favors to the largest extent the grain size refinement of TiNi alloys during milling. The effect of chemical modification of TiNi alloy by Mg element is not clear. Some researchers claim that Ni replacement by Mg improves the discharge capacity and cycle life the electrodes prepared from TiNi alloy [17]. On the other hand, another group argues that the discharge capacity of TiNi alloy decrease monotonically with increasing Mg content [20]. Moreover, addition of Mg to TiNi phase has hindered the amorphization progress as all phases incline toward crystallization with the increase of Mg content [20].

Ti–Ni-based alloys were also used in the past to improve electrochemical properties of other alloys. For example, ball milling was used to produce TiNi-Mg$_2$Ni alloy. Cycle life of the electrode was improved. It was concluded that TiNi particles are the active sites for the redox reaction of the hydrogen atoms and the pathway for the diffusion of hydrogen. It was also found that introduction of TiNiCo alloy could significantly improve the catalytic activation of the electrode and decrease the charge transfer resistance and the diffusion impedance of the hydrogen atoms [42].

10.5 Other Methods of Alloy Production

Besides the described methods (arc melting, MA), Ti–Ni materials can be produced by other methods, too. Recently, Ti$_2$Ni was synthesized by the electro-deoxidation technique. With this method, alloys can be obtained directly from their oxides in molten salts. It provides great economical advantage, especially if the multi-component alloy synthesis can be achieved [53]. Ti$_2$Ni alloy was also synthesized in the molten CaCl$_2$ electrolyte by the electro-deoxidation method at 900°C [26].

Moreover, the arc spraying technique was also used to obtain TiNi alloy skeleton electrodes [54]. TiNi skeleton electrodes show much higher activity for the hydrogen evolution reaction (HER) than skeleton Ni electrode. The catalytic effect of Ti significantly increases the real exchange current density and confers an enhanced activity, despite the large value of the Tafel slope. Also, it reduces the electrochemical activation energy for the HER [54].

Ti–Ni-based alloys were also prepared by rapid quenching of the metal and treatment with HF [30].

10.6 Gaseous Hydrogen Sorption and Desorption of Ti–Ni Alloys

Ti–Ni-based materials were much less studied for gaseous hydrogenation properties than for electrochemical properties. In the past, Takeshita et al. studied Ti$_2$Ni-based alloys. This group showed that Ti$_2$Ni alloy can reversibly absorb and desorb hydrogen under moderate conditions (room temperature and atmospheric pressure). The PCT relation exhibited a hydrogen pressure plateau (for modified alloys) and insignificant hysteresis [42]. The plateau pressure is attributed to α-metal to β-hydride phase transformation [21]. Ti$_2$Ni compounds modified by O, N, and C demonstrated higher hydrogen desorption pressure than unmodified Ti$_2$Ni alloy [42]. Oxygen addition to the binary Ti$_2$Ni alloy leads to an increase in the hydrogen desorption pressure of the alloy. Oxygen, nitrogen, and carbon atoms occupy the 16c positions in the Ti$_2$Ni-based compounds. All of chemical modifiers decreased the stability of hydrogen occlusion phases of the Ti$_2$Ni-based material [55]. There was a decrease in the hydrogen desorption temperature with increased O-content [3]. However, Saldan et al. reported that with an increase in the amount of added oxygen, the amount of the absorbed hydrogen decreases, which is reflected in the decreased volumes of the crystal lattice for all hydrides [10, 13].

Partial substitution of Ti by Zr in the Ti–Ni alloy leads to a substantial increase of the hydrogenation capacity with simultaneous improvement of absorption kinetic. All of these studied materials were characterized by the same h-phase type structure. All alloys actively absorb hydrogen at ambient temperature giving saturated hydrides over 20–60 min [3].

Also other additives influence the hydrogen sorption properties: The pressure of plateau increases with Cu content showing that the hydride phase destabilizes with Cu substitution [21]; partial substitution of Ni by Fe leads to better hydrogen storage capacity [10].

10.7 Electrochemical and Gaseous Hydrogen Sorption and Desorption of Ti$_2$Ni Chemically Modified by Pd and Multi-Walled Carbon Nanotubes

Recently, Ti$_2$Ni alloy which was produced by mechanical alloying and heat treatment was chemically modified by multi-walled carbon nanotubes (MWCNTs) and Pd in order to enhance its electrochemical and gaseous hydrogen storage properties. Ti$_2$Ni and Ti$_2$Ni + 5 wt% Pd alloys were prepared first by the MA process to obtain amorphous phase in 8 h of milling. Amorphous materials were then annealed at 750°C for 0.5 h. Heat treatment resulted in obtaining of a Ti$_2$Ni cubic structure with TiNi minor phase (in case of Ti$_2$Ni + 5 wt%Pd alloy) . As shown by XRD measurements, addition of Pd caused a slight growth of lattice constant and cell volume. This could be because Pd has larger atomic size than Ni. Moreover, chemically modified Pd alloy was characterized by bigger average crystallites size (about 42 nm) [15, 43].

MWCNTs together with theabove-mentioned crystalline materials were used to prepare nanocomposites. They were prepared by ball co-milling of MWCNTs with some Ti$_2$Ni-type alloy. XRD measurements revealed that co-milling for longer time led to the amorphization of nanocomposites. AFM and TEM micrographs confirmed that MWCNTs are well distributed on boundaries and surfaces of agglomerates after 1 min of co-milling (Fig. 10.6). Longer milling time led to the destruction of MWCNTs. As an effect of MWCNTs addition, alloy particles have a smaller size and are characterized by less agglomeration and better dispersion due to the their lubricant function. As electrochemical measurements have shown, longer milling time led to lower discharge capacity due to the amorphization of the alloy and decay of MWCNTs [15, 43].

From all the measured materials, Ti$_2$Ni + 5 wt% Pd + 5 wt% MWCNTs is the material with the highest discharge capacity— 301 mAh/g [15]. Figure 10.7 shows discharge capacity as a function of cycle number of Ti$_2$Ni electrode and Ti$_2$Ni + Pd + MWCNTs nanocomposite electrode. Obtained high discharge capacity is due to

- applied method which provides obtaining of nanocrystalline materials;
- enlargement of cell volume due to Pd addition;
- improvement of charge-transfer reaction, due to better electrochemical catalytic activity for charge-transfer reaction of Pd than Ni;
- increased specific surface area and a smaller size of particles which shorten diffusion distance of hydrogen to inner part of alloy due to the lubricant function of MWCNTs;
- good channels for hydrogen transport, provided by tubular structure of MWCNTs.

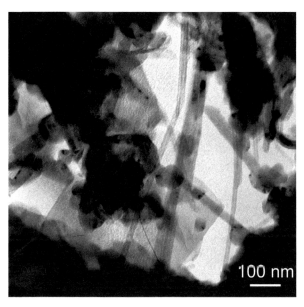

Figure 10.6 TEM micrograph of $Ti_2Ni+Pd+MWCNT$ nanocomposite.

The purposes of capacity loss during cycling were listed and discussed in Section 10.6.

Ti_2Ni-based alloys and nanocomposites were also tested for gaseous hydrogen sorption properties. PCI absorption curves show that all of tested materials actively absorb hydrogen at ambient temperature and at about 1 MPa. Measurements show that addition of MWCNTs causes obtaining of material with better activation properties. The lubricant effect of MWCNTs reduces the stickiness of powders and consequently reduces the particle size and increases the specific surface area. The highest capacity among tested materials was obtained for Ti$_2$Ni + Pd—2.1 wt%. Moreover, chemical modification by Pd affects the kinetic of hydrogen sorption, too, greatly reducing the sorption time [3].

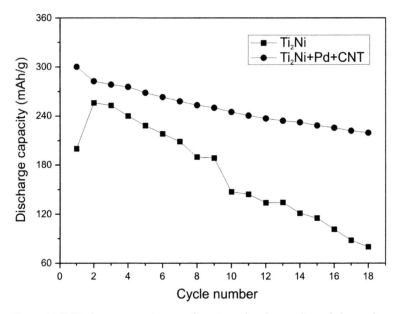

Figure 10.7 Discharge capacity as a function of cycle number of electrodes prepared with Ti$_2$Ni alloy and Ti$_2$Ni + Pd + MWCNTs nanocomposite.

References

1. Balcerzak, M., and Jurczyk, M. (2015). Influence of gaseous activation on hydrogen sorption properties of TiNi and Ti$_2$Ni alloys, *J. Mater. Eng. Perform.* **24**: 1710–1717.

2. Krstajic, N. V., Grgur, B. N., Mladenovic, N. S., Vojnovic, M. V., and Jaksic, M. M. (1997). The determination of kinetics parameters of the hydrogen evolution on Ti–Ni alloys by ac impedance, *Electrochem. Acta* **42**: 323–330.

3. Zavaliy, I., Wojcik, G., Mlynarek, G., Saldan, I., Yartys, V., and Kopczyk, M. (2001). Phase-structure characteristics of (Ti$_{1-x}$Zr$_x$)$_4$Ni$_2$O$_{0.3}$ alloys and their hydrogen gas and electrochemical absorption-desorption properties, *J. Alloy Comp.*, **314**: 124–131.

4. Yang, M., Zhao, X., Ding, Y., Ma, L., Qu, X., and Gao, Y. (2010). Electrochemical properties of titanium-based hydrogen storage alloy prepared by solid phase sintering, *Int. J. Hydrogen Energy*, **35**: 2717–2721.
5. Zhao, X., Ma, L., Gao, Y., and Shen, X. (2009). Structure, morphology and hydrogen desorption characteristics of amorphous and crystalline Ti–Ni alloys, *Mater. Sci. Eng. A* **516**: 50–53.
6. Cui, N., Luan, B., Zhao, H. J., Liu, H. K., and Dou, S. X. (1996). Synthesis and electrode characteristics of the new composite alloys Mg$_2$Ni-x wt%Ti$_2$Ni, *J. Alloys Comp.* **240**: 229–234.
7. Toprek, D., Belosevic-Cavor, J., and Koteski, V. (2015). Ab initio studies of the structural, elastic, electronic and thermal properties of NiTi$_2$ intermetallic, *J. Phys. Chem. Sol.* **85**: 197–205.
8. Massalski, T. B., Okamoto, H., Subramanian, P. R., and Kacprzak, L. (1990). *Binary Alloy Chase Diagrams*, 2nd ed., vol. 3. Materials Park, OH: ASM International, p. 2874.
9. Luan, B., Kennedy, S. J., Liu, H. K., and Dou, S. X. (1998). On the charge/discharge behavior of Ti$_2$Ni electrode in 6 M KOH aqueous and deuterium oxide solution, *J. Alloy Comp.* **267**: 224–230.
10. Saldan, I., Burtovyy, R., Becker, H. W., Ader, V., and Woll, Ch. (2008). Ti–Ni alloys as MH electrodes in Ni-MH accumulators, *Int. J. Hydrogen Energy*, **33**: 7177–7184.
11. Zhao, X., Ma, L., Ding, Y., and Shen, X. (2010). Structural evolution and electrochemical hydrogenation behavior of Ti$_2$Ni alloy, *Intermetallics*, **18**: 1086–1090.
12. Luan, B., Cui, N., Zhao, H., Liu, H. K., and Dou, S. X. (1995). Mechanism of early capacity loss of Ti$_2$Ni hydrogen storage alloy electrode, *J. Power Sources*, **55**: 101–106.
13. Saldam, I. V., Koval'chuk, I. V., and Zavalii, I. Y. (2003). Influence of oxygen modification and alloying on the charge-discharge characteristics of metal-hydride electrodes based on Ti$_2$Ni, *Mater. Sci.* **39**: 545–553.
14. Zhao, X., Zhou, J., Shen, X., Yang, M., and Ma, L. (2012). Structure and electrochemical hydrogen storage properties of A$_2$B-type Ti–Zr–Ni alloy, *Int. J. Hydrogen Energy*, **37**: 5050–5055.
15. Balcerzak, M., Nowak, M., Jakubowicz, J., and Jurczyk, M. (2014). Electrochemical behavior of nanocrystalline TiNi doped by MWCNTs and Pd, *Renew. Energy*, **62**: 432–438.

16. Drenchev, B., and Spassov, T. (2007). Electrochemical hydriding of amorphous and nanocrystalline TiNi-based alloys, *J. Alloys Comp.* **441**: 197–201.

17. Drenchev, B., Spassov, T., and Radev, D. (2008). Influence of alloying and microstructure on the electrochemical hydriding of TiNi-based ternary alloys, *J. Appl. Electrochem.*, **38**: 437–444.

18. Jankowska, E., Makowiecka, M., and Jurczyk, M. (2008). Electrochemical performance of sealed Ni-MH batteries using nanocrystalline TiNi-type hydride electrodes, *Renew. Energy*, **33**: 211–215.

19. Balcerzak, M., Nowak, M., Jakubowicz, J., and Jurczyk, M. (2014). Electrochemical behavior of nanocrystalline TiNi doped by MWCNTs and Pd, *Renew. Energy*, **62**: 432–438.

20. Li, X. D., Elkedim, O., Nowak, M., and Jurczyk, M. (2014). Characterization and first principle study of ball milled Ti–Ni with Mg doping as hydrogen storage alloy, *Int. J. Hydrogen Energy* **39**: 9735–9743.

21. Emami, H., Cuevas, F., and Latroche, M. (2014). Ti(Ni,Cu) pseudobinary compounds as efficient negative electrodes for Ni-MH batteries, *J. Power Sources*, **265**: 182–191.

22. Emami, H., Souques, R., Crivello, J.-C., and Cuevas, F. (2013). Electronic and structural influence of Ni by Pd substitution on the hydrogenation properties of TiNi, *J. Solid State Chem.*, **198**: 475–484.

23. Luan, B., Zhao, H., Liu, H. K., and Dou, S. X. (1996). On discharge process of titanium-based hydrogen storage alloy electrode via a.c. impedance analysis, *J. Power Sources*, **62**: 75–79.

24. Wang, C. S., Lei, Y. Q., and Wang, Q. D. (1998). Studies of electrochemical properties of TiNi alloy used as an MH electrode—I. Discharge capacity, *Electrochem. Acta*, **43**: 3193–3207.

25. Guiose, B., Cuevas, F., Decamps, B., Leroy, E., and Percheron-Guegan, A. (2009). Microstructural analysis of the ageing of pseudo-binary (Ti,Zr)Ni intermetallic compounds as negative electrodes of Ni-MH batteries, *Electrochim. Acta*, **54**: 2781–2789.

26. Hosni, B., Li, X., Khaldi, C., Elkedim, O., and Lamloumi, J. (2014). Structure and electrochemical hydrogen storage properties of Ti_2Ni alloy synthesized by ball milling, *J. Alloy Comp.* **615**: 119–125.

27. Drenchev, B., and Spassov, T. (2009). Influence of B substitution for Ti and Ni on the electrochemical hydriding of TiNi, *J. Alloys Comp.* **474**: 527–530.

28. Xu, Y.-H., Chen, C.-P., Wang, Q.-D., and Chen, L.-X. (2001). The effect of temperature on the electrode properties of the directly prepared Ti_3Ni_2 alloy, *Int. J. Hydrogen Energy*, **26**: 1177–1181.

29. Xu, Y.-H., Chen, C. P., and Wang, Q.-D. (2001). The influence of small amount of added elements on the electrode performance characteristics for Ti$_3$Ni$_2$ hydrogen storage alloy, *Mater. Chem. Phys*, **71**: 190–194.
30. Krstajic, N. V., Grgur, B. N., Zdujic, M., Vojnovic, M. V., and Jaksic, M. M. (1997). Kinetic properties of the Ti–Ni intermetallic phases and alloys for hydrogen evolution, *J. Alloy Comp.* **257**: 245–252.
31. Saldan, I. V., Dubov, Y. H., Ryabov, O. B., and Zavalii, I. Y. (2006). Effect of the modification of metal-hydride electrodes based on Ti$_2$Ni alloys on their discharge characteristics, *Mater. Sci.* **42**: 634–642.
32. Wang, C. S., Lei, Y. Q., and Wang, Q. D. (1998). Effects of Nb and Pd on the electrochemical properties of a Ti–Ni hydrogen-storage electrode, *J. Power Sources* **70**: 222–227.
33. Luan, B., Cui, N., Zhao, H., Zhong, S., Liu, H. K., and Dou, S. X. (1996). Studies on the performance of Ti$_2$Ni$_{1-x}$Al$_x$ hydrogen storage alloy electrodes, *J. Alloy Comp.* **233**: 225–230.
34. Zavalii, I. Y., and Saldan, I. V. (2002). Investigation of Ti(Zr-Ni) hydrogen-sorbing alloy as electrode materials for nickel-metal-hydride storage batteries, *Mater. Sci.* **38**: 526–533.
35. Zhao, X., Li, J., Yao, Y., Ma, L., and Shen, X. (2012). Electrochemical hydrogen storage properties of a non-equilibrium Ti$_2$Ni alloy, *RSC Adv.*, **2**: 2149–2153.
36. Luan, B., Cui, N., Liu, H. K., Zhao, H., and Dou, S. X. (1994). Low-temperature surface micro-encapsulation of Ti$_2$Ni hydrogen-storage alloy powder, *J. Power Sources*, **52**: 295–299.
37. Zhao, X., Ma, L., Qu, X., Ding, Y., and Shen, X. (2009). Effect of Mechanical Milling on the structure and electrochemical properties of Ti$_2$Ni alloy in an alkaline battery, *Energy Fuels*, **23**: 4678–4682.
38. Luan, B., Cui, N., Liu, H. K., Zhao, H. J., and Dou, S. X. (1995). Effect of cobalt addition on the performance of titanium-based hydrogen-storage electrodes, *J. Power Sources*, **55**: 197–203.
39. Zhao, X., Ma, L., Yao, Y., Ding, Y., and Shen, X. (2010). Ti$_2$Ni alloy: A potential candidate for hydrogen storage in nickel/metal hydride secondary batteries, *Energy Environ. Sci.*, **3**: 1316–1321.
40. Zhao, X., Ma, L., Yang, M., Ding, Y., and Shen, X. (2010). Electrochemical properties of Ti–Ni–H powders prepared by milling titanium hydride and nickel, *Int. J. Hydrogen Energy*, **35**: 3076–3079.
41. Wen, H., Jianli, W., Lidong, W., Yaoming, W., and Limin, W. (2009). Electrochemical hydrogen storage in (Ti$_{1-x}$V$_x$)$_2$Ni (x = 0.05–0.3) alloys

comprising icosahedral quasicrystalline phase, *Electrochim. Acta*, **54**: 2770–2773.

42. Zlatanova, Z., Spassov, T., Eggeler, G., and Spassova, M. (2011). Synthesis and hydriding/dehydriding properties of Mg$_2$Ni-AB (AB = TiNi or TiFe) nanocomposites, *Int. J. Hydrogen Energy* **36**: 7559–7566.

43. Balcerzak, M., Jakubowicz, J., Kachlicki, T., and Jurczyk, M. (2015). Hydrogenation properties of nanostuctured Ti$_2$Ni-based alloys and nanocomposites, *J. Power Sources* **280**: 435–445.

44. Balcerzak, M., and Jurczyk, M. (2014). Electrochemical and corrosion behavior of nanocrystalline TiNi-based alloys and composite, *Acta Phys. Polonica A*, **126**: 888–891.

45. Balcerzak, M., and Jurczyk, M. (2014). The influence of chemical modification by silver on hydrogen storage properties of nanocrystalline Ti$_2$Ni alloy, *Acta Phys. Polonica A*, **126**: 892–894.

46. Vaivars, G., Kleparis, J., Mlynarek, G., Wojcik, G., and Zavaliy, I. (1999). AC impedance behavior of the Ti$_4$Ni$_2$O$_y$ and Ti$_{3.5}$Zr$_{0.5}$Ni$_2$O$_y$ type metal hydride electrodes, *Ionics*, **5**: 292–298.

47. Takeshita, H. T., Tanaka, H., Kiyobayashi, T., Takeichi, N., and Kuriyama, N. (2002). Hydrogenation characteristics of Ti$_2$Ni and Ti$_4$Ni$_2$X (X = O, N, C), *J. Alloy Comp.* **330–332**: 517–521.

48. Sheppard, D. A., Jiang, Z. T., and Backley, C. E. (2007). Investigations of hydrogen uptake in ball-milled TiMgNi, *Int. J. Hydrogen Energy* **32**: 1928–1932.

49. Jian, L., Xueping, G., Deying, S., Yunshi, Z., and Shihai, Y. (1995). The characteristics of the microencapsulated Ti–Ni alloys and their electrodes, *J. Alloy Comp.* **231**: 852–855.

50. Luan, B., Liu, H. K., and Dou, S. X. (1997). On the elemental substitution of titanium-based hydrogen-storage alloy electrodes for rechargeable Ni-MH batteries, *J. Mater. Sci.* **32**: 2629–2635.

51. Li, X. D., Elkedim, O., Nowak, M., Jurczyk, M., and Chassagnon, R. (2013). Structural characterization and electrochemical hydrogen storage properties of Ti$_{2-x}$Zr$_x$Ni (x = 0, 0.1, 0.2) alloys prepared by mechanical alloying, *Int. J. Hydrogen Energy* **38**: 12126–12132.

52. Szajek, A., Makowiecka, M., Jankowska, E., and Jurczyk, M. (2005). Electrochemical and electronic properties of nanocrystalline TiNi$_{1-x}$M$_x$ (M = Mg, Mn, Zr; x = 0, 0.125, 0.25) ternary alloys, *J. Alloy Comp.* **403**: 323–328.

53. Anik, M., Baksan, B., Orbay, T. O., Kucukdeveci, N., Aybar, A. B., Ozden, R. C., Gasan, H., and Koc, N. (2014). Hydrogen storage characteristics

of Ti_2Ni alloy synthesized by the electro-deoxidation technique, *Intermetallics* **46**: 51–55.

54. Kellenberger, A., Vaszilcsin, N., Brandl, W., and Duteanu, N. (2007). Kinetics of hydrogen evolution reaction on skeleton nickel and nickel-titanium electrodes obtained by thermal arc spraying technique, *Int. J. Hydrogen Energy* **32**: 3258–3265.

55. Takeshita, H. T., Tanaka, H., Kuriyama, N., Sakai, T., Uehara, I., and Haruta, M. (2000). Hydrogenation characteristics of ternary alloys containing Ti_4Ni_2X (X = O, N, C), *J. Alloy Comp.* **311**: 188–193.

Chapter 11

ZrV$_2$-Based Hydrogen Storage Alloys

Marek Nowak and Mieczyslaw Jurczyk

*Institute of Materials Science and Engineering,
Poznan University of Technology, Poznan, Poland*

marek.nowak@put.poznan.pl, mieczyslaw.jurczyk@put.poznan.pl

11.1 Zr–V Phase Diagram and Structure

There is much interest in Zr-based AB$_2$-type alloys, as many exhibit desirable electrochemical properties and are among the most promising materials for hydrogen energy applications, such as hydrogen storage materials, heat pumps, and electrode materials for rechargeable nickel-metal hydride (Ni-MH$_x$) batteries [1–16].

The assessed phase diagram for the Zr–V system is based primarily on E. Rudy's investigation of the phase relationships (Fig. 11.1) [17, 18]. Williams indicated negligible solubility of V in Zr and a negligible range of homogeneity for the intermediate ZrV$_2$ phase [19]. Below room temperature, ZrV$_2$ has been widely studied because a martensitic transition has been reported in the range 173 to –143°C [20], and a superconducting transition temperatures in the range –266 to –264°C [21].

Handbook of Nanomaterials for Hydrogen Storage
Edited by Mieczyslaw Jurczyk
Copyright © 2018 Pan Stanford Publishing Pte. Ltd.
ISBN 978-981-4745-66-6 (Hardcover), 978-1-315-36444-5 (eBook)
www.panstanford.com

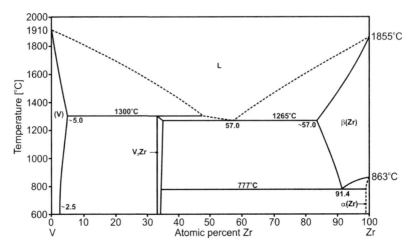

Figure 11.1 Zr–V phase diagram [17].

Laves phases are a group of three topologically highly closed-packed inter-metallic compounds with the stoichiometry of AB_2: C15 with a face-center-cubic $MgCu_2$ structure, C14 with a hexagonal $MgZn_2$ structure, and C36 with a dihexagonal $MgNi_2$ structure [17, 22]. Most of the research focus on the C14 (ZrV_2) and C15 ($Zr_{0.35}Ti_{0.65}V_{0.85}Cr_{0.26}Ni_{1.30}$) phases. The C36 phase may be an intermediate phase between C14 and C15 [22–24]. The main controlling factor for the C14/C15 ratio is the average electron density (e/a), geometric atomic size ratio, and difference in electro negativity [25].

The e/a value is calculated by averaging the numbers of outer-shell electrons of all constituent elements in the formulation. It has been reported that based on the Laves phase structures with $e/a > 5.76$, the C15 structure is stabilized. At an e/a range of 5.88 to 7.53, the C14 structure is stabilized and finally with $e/a > 7.65$, the C15 structure is stabilized again [26].

Another theory in controlling the preference of C14/C15 structures is geometric atomic size ratio. The Laves phases occur between r_A/r_B ratio range of ~1.05 to 1.68, where r_A and r_B are the average metallic radii of A atoms and B atoms, respectively [27]. Within a reasonable r_A/r_B ratio range, the atoms are able to adjust themselves by expanding or contracting to accommodate the ideal r_A/r_B ratio during the Laves phase formation [28]. As a confirmation of the ideal atomic size ratio and r_A/r_B ratio range for the Laves phases, a study where the enthalpies of formation (ΔH_f) of many Laves phases were plotted as a function of the r_A/r_B ratio has shown that the lowest ΔH_f (most stable) is at an r_A/r_B ratio around 1.225 and lower and higher bounds intersected with $\Delta H_f = 0$ (the compound is thermodynamically unstable beyond this point) are at r_A/r_B ratios of 1.03 and 1.65, respectively [29, 30]. By organizing the details of the existing binary and ternary Laves phases, it was shown that the higher solubility ranges of C14 and C36 phases are observed most frequently within the r_A/r_B ratio range of 1.12 to 1.26, and the higher solubility range of C15 phase is typically seen within the r_A/r_B ratio range of 1.1 to 1.35 [31].

11.2 ZrV$_2$ Type Alloys Synthesized by Mechanical Alloying

The ZrV$_2$ alloy crystallizes in the cubic C15-type structure. Nevertheless, the application of this type of materials in batteries has been limited due to poor absorption/desorption kinetics in addition to a complicated activation procedure. While many studies on AB$_2$ alloys were reported, most of them focused on the influences of A- or B-site substitutions on electrochemical properties; for example, $Zr_{0.35}Ti_{0.65}V_{0.85}Cr_{0.26}Ni_{1.30}$ alloy. These results are summarized in Table 11.1.

It has been shown, that nanocrystalline ZrV$_2$ and $Zr_{0.35}Ti_{0.65}V_{0.85}Cr_{0.26}Ni_{1.30}$ alloys can be synthesized by mechanical alloying followed by annealing [32–36]. The electrochemical properties of nanocrystalline powders were measured and compared with those of amorphous material. The behavior of the MA process has been studied by X-ray diffraction. The Zr + 2V powder mixture milled for more than 25 h has transformed absolutely to the amorphous phase (Fig. 11.2). Formation of the ordered C14 (ZrV$_2$) phase was achieved by annealing the amorphous materials in high purity argon atmosphere at 800°C for 0.5 h (Fig. 11.2d).

Table 11.1 Summary of recent research on Laves AB_2 phase [4, 5, 10]

Substitution	Major effect
Zr	Forms stable hydrides ($\Delta H = -82$ kJ/mol for ZrH_2). Absorber of gaseous O_2, N_2, CO_2, etc. Oxidizes quickly, forming compact passive oxide layer; increasingly loose when alloyed with Ti. High electro catalytic activity
Ti	Forms stable hydrides ($\Delta H = -68$ kJ/mol for TiH_2). Complete solubility with Zr: Inhibits Zr oxidation. Oxidizes, forming loose passive oxide layer. Dissolution in KOH. When Ti replaces for Zr in AB_2 alloy, the plateau pressure of hydrogen increases, hydrogen absorption capacity becomes smaller, but hysteresis in the PC isotherms decreases
V	Forms stable hydrides ($\Delta H = -35$ kJ/mol for $VH_{0.5}$). On B site in AB_2 structure, improves hydrogen sorption capacity and kinetics. When on A site in AB_2, the alloy hardly absorbs hydrogen. Both in AB_5 and AB_2 alloys forms oxides soluble in alkaline solution. Increases the alloy surface owing to dissolution in alkaline solution
Cr	Forms unstable hydride ($\Delta H = -6$ kJ/mol for CrH, $\Delta H = -8$ kJ/mol for $CrH_{0.5}$). Inhibitor of vanadium corrosion. Influences on M-H bond strength in alloy. Good cycling properties. Easily forms $TiCr_2$ and separate bcc phase with vanadium. Is not easily introduced in C14 phase
Ni	Forms unstable hydride ($\Delta H = -3$ kJ/mol for $NiH_{0.5}$). Indispensable element because of its high electrocatalytic activity. Forms intermetallics, decreases the Me-H bond strength to a suitable level. Sensitive to corrosion and oxidation during cycling [forms $Ni(OH)_2$]. Decreases oxidation of Zr-Ti-V alloys and increases the discharge capacity. Excessively high Ni content leads to decrease in the discharge capacity

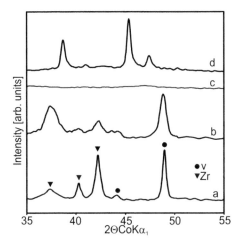

Figure 11.2 XRD spectra of a mixture of Zr and V powders mechanically alloyed for different times in an argon atmosphere: (a) initial state (elemental powder mixture), (b) after MA for 5 h, (c) after MA for 25 h and (d) heat treated at 800°C for 0.5.

Single phase formation of $Zr_{0.35}Ti_{0.65}V_{0.85}Cr_{0.26}Ni_{1.30}$ alloy with C15 structure by mechanical alloying and annealing was presented in Fig. 11.3.

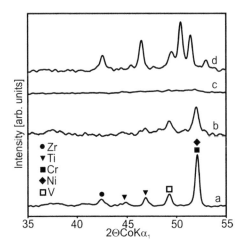

Figure 11.3 XRD spectra of a mixture of Zr, Ti, V, Cr and Ni powders mechanically alloyed for different times in an argon atmosphere: (a) initial state (elemental powder mixture), (b) after MA for 5 h, (c) after MA for 25 h and (d) heat treated at 800°C for 0.5 h.

According to AFM studies, the average size of annealed Zr–V powders was of order of 23 nm (Fig. 11.4). Proper engineering of microstructure and surface by using unconventional processing techniques, such as mechanical alloying (MA) [33, 37–40] or high-energy ball-milling (HEBM) [32, 34, 37, 41, 42], can lead to advanced nanocrystalline intermetallics representing a new generation of metal hydride electrodes. These materials show substantially enhanced absorption characteristics superior to that of the conventionally prepared materials [41]. Zaluski et al. reported also, that powders of nanocrystalline alloys, modified with a catalyst, readily absorb hydrogen, with no need for prior activation [41]. The generation of new metastable phases or materials with an amorphous grain boundary phase offers a wider distribution of available sites for hydrogen and thus a totally different hydrogenation behavior.

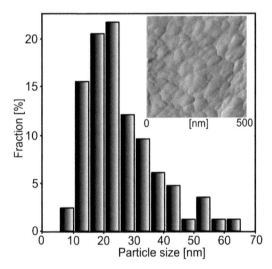

Figure 11.4 AFM picture and particle size dimension distribution histograms of ZrV_2 powdered sample (30 h MA, heat treated at 800°C for 0.5 h).

11.3 Electrochemical and Thermodynamic Properties

To study the quality of the activated microcrystalline, amorphous and nanocrystalline $Zr_{0.35}Ti_{0.65}V_{0.85}Cr_{0.26}Ni_{1.30}$, as electrode materials in the Ni-MH$_x$ battery, the overpotential against current density was recorded at 15 s of anodic and cathodic galvanostatic pulses (Fig. 11.5). It can be seen that the anodic and cathodic parts of microcrystalline and nanocrystalline electrodes are almost symmetrical with respect to the rest potential of electrode. It may be concluded that for both electrodes fast discharge rates can be achieved.

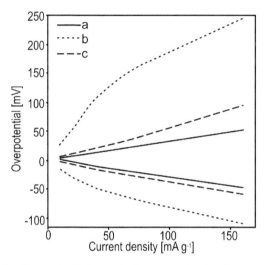

Figure 11.5 Overpotential against current density on activated (a) microcrystalline, (b) amorphous, and (c) nanocrystalline $Zr_{0.35}Ti_{0.65}V_{0.85}Cr_{0.26}Ni_{1.30}$ electrodes, at 15 s of anodic and cathodic galvanostatic pulses.

The electrochemical pressure composition (e.p.c.) isotherms for absorption and desorption of hydrogen were obtained from the equilibrium potential values of the electrodes, measured during intermittent charge and/or discharge cycles at constant current density, by using the Nernst equation. The e.p.c. isotherms determined on the studied materials are illustrated in Fig. 11.6. Due to the amorphous nature of the $Zr_{0.35}Ti_{0.65}V_{0.85}Cr_{0.26}Ni_{1.30}$ alloy prior to annealing (curve (b) on Fig. 11.6), the hydrogen absorption-desorption characteristics are not satisfactory. Annealing causes transformation from the amorphous to the crystalline structure and produces grain boundaries. Anani et al. [3] noted that grain boundaries are necessary for the migration of the hydrogen into the alloy. It is worth noting from Fig. 11.6 that the characteristics for microcrystalline and nanocrystalline materials are very similar in respect to hydrogen contents, but there are small differences in the plateau pressures.

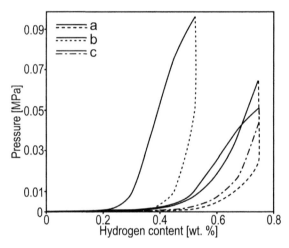

Figure 11.6 Electrochemical pressure–composition isotherm for absorption (solid line) and desorption (dashed line) of hydrogen on (a) microcrystalline, (b) amorphous, and (c) nanocrystalline $Zr_{0.35}Ti_{0.65}V_{0.85}Cr_{0.26}Ni_{1.30}$, at 20°C.

For comparison, Fig. 11.7 shows the storage capacity of microcrystalline $Zr_{0.35}Ti_{0.65}V_{0.85}Cr_{0.26}Ni_{1.30}$ alloy for hydrogen absorption and desorption from the gas phase, as characterized by equilibrium hydrogen pressure against hydrogen content isotherms (p.c.) measured using Sievert equipment. It is worth noting that the electrochemical capacity of the microcrystalline alloy electrode is smaller (see Fig. 11.6a). These results point to differences in the surface state of alloy particles in the electrode material in an alkaline solution and in the hydrogen gas atmosphere. Generally, the electrochemical oxidation of particle surface in an alkaline solution, along with formation of a passive layer, inhibits the hydrogen electrosorption into the studied microcrystalline electrode material. This is believed to be induced by the formation of oxide layer (TiO_2) on the electrode surface during the cycle test, which prevented hydrogen from penetrating the surface of electrode [44].

The typical galvanostatic discharge curves of the electrodes fabricated from the microcrystalline, amorphous and nanocrystalline

$Zr_{0.35}Ti_{0.65}V_{0.85}Cr_{0.26}Ni_{1.30}$ materials are shown in Fig. 11.8. It can be seen that the discharge capacities of microcrystalline as well nanocrystalline electrodes are almost the same. As shown in Fig. 11.8, the galvanostatic discharge characteristics of the Ni-MH$_x$ battery with an amorphous material as the negative electrode is clearly lower.

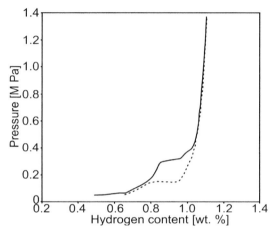

Figure 11.7 Pressure composition isotherm for absorption (solid line) and desorption (dashed line) of hydrogen from the gas phase on microcrystalline $Zr_{0.35}Ti_{0.65}V_{0.85}Cr_{0.26}Ni_{1.30}$, at 20°C.

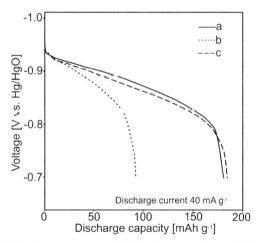

Figure 11.8 Discharge curves of (a) microcrystalline, (b) amorphous and (c) nanocrystalline $Zr_{0.35}Ti_{0.65}V_{0.85}Cr_{0.26}Ni_{1.30}$ electrodes, at the seventh cycle (current density of 40 mA g^{-1} in 6M KOH).

The C14 and C15 phases have different contributions to the gaseous phase hydrogen storage and electrochemical properties of the AB$_2$ alloys. The C15 phase has both superior gaseous phase hydrogen storage capacity and reversibility, a better high rate discharge ability (HRD) due to better reaction in the bulk, a greater specific power, and an improved low temperature performance with a shortcoming of an inferior cycle life when compared to the C14 phase [45].

Figure 11.9 shows the discharge capacities of the electrodes as a function of charge-discharge cycling number. The electrode prepared with nanocrystalline $Zr_{0.35}Ti_{0.65}V_{0.85}Cr_{0.26}Ni_{1.30}$ material showed better activation and higher discharge capacities. This improvement is due to a well-established diffusion path for hydrogen atoms along the numerous grain boundaries [46]. It is worth noting, that annealed nanocrystalline powder has greater capacities than the amorphous parent alloy powders. Table 11.2 reports the discharge capacities of the studied materials. The electrochemical results show very little difference between the nanocrystalline and microcrystalline powders, compared with the substantial difference between these and the amorphous powder. In the annealed nanocrystalline $Zr_{0.35}Ti_{0.65}V_{0.85}Cr_{0.26}Ni_{1.30}$ powders prepared by mechanical alloying and annealing, discharging capacities up to 150 mA h g^{-1} (at 160 mA g^{-1} discharge current) have been measured [32].

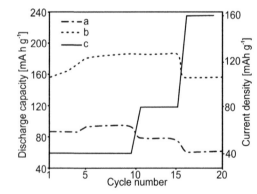

Figure 11.9 Discharge capacity as a function of cycle numbers of electrode prepared with (a) amorphous and (b) nanocrystalline $Zr_{0.35}Ti_{0.65}V_{0.85}Cr_{0.26}Ni_{1.30}$ (solution, 6M KOH; temperature, 20°C). The charge conditions were 40 mA g^{-1}; The discharge conditions were plotted in the figure (c); cut-off potential vs. Hg/HgO/6M KOH was −0.7 V.

Table 11.2 Discharge capacities of microcrystalline, amorphous and nanocrystalline $Zr_{0.35}Ti_{0.65}V_{0.85}Cr_{0.26}Ni_{1.30}$ materials (current density of charging and discharging was 160 mA g^{-1})

Preparation method	Structure type	Discharge capacity at 18th cycle [mA h g^{-1}]
arc melting and annealing*	microcrystalline (MgZn$_2$)	135
MA	amorphous	65
MA and annealing	nanocrystalline (MgZn$_2$)	150

*1000°C/7 days.

Independently, it has been shown that the electrochemical properties of hydrogen storage alloys, which do not contain nickel, can be stimulated by high-energy ball-milling of the precursor alloys with a small amount of nickel powders [47, 48]. The ZrV_2 and $Zr_{0.5}Ti_{0.5}V_{0.8}Mn_{0.8}Cr_{0.4}$ alloy powders have been prepared using HEBM of precursor alloys with nickel powder. It was confirmed that discharge capacity of electrodes prepared with application of ZrV_2 and $Zr_{0.5}Ti_{0.5}V_{0.8}Mn_{0.8}Cr_{0.4}$ alloy powders with 10 wt% of nickel powder addition was impossible to estimate because of extremely high polarization. In the electrodes prepared with application of high-energy ball-milled ZrV_2/Ni and $Zr_{0.5}Ti_{0.5}V_{0.8}Mn_{0.8}Cr_{0.4}$/Ni alloy powders with 10 wt% of nickel powder the discharge capacities were considerably improved, and they increased from 0 to 110 mA h g^{-1} and 214 mA h g^{-1}, respectively (Table 11.3).

Table 11.3 Discharge capacities of nanocrystalline ZrV_2-type materials without and with 10 wt% of Ni powder (current density of charging and discharging was 4 mA g^{-1})

Composition	Discharge capacity [mA h g^{-1}]
ZrV_2	0
ZrV_2/Ni	110
$Zr_{0.5}Ti_{0.5}V_{0.8}Mn_{0.8}Cr_{0.4}$	0
$Zr_{0.5}Ti_{0.5}V_{0.8}Mn_{0.8}Cr_{0.4}$/Ni	214

The alloy elements such as Ti substituted for Zr and Mn, Cr substituted for V in ZrV_2/Ni-based materials greatly improved the activation behavior of the electrodes. It is worth noting that annealed nanocrystalline ZrV$_2$/Ni-based powders have greater capacities (about 2.2 times) than the amorphous parent alloy powders. Generally, the electrochemical properties are closely linked to the size and crystallographic perfection of the constituent grains, which in turn are a function of the processing or grain refinement method used to prepare the hydrogen storage alloys.

11.4 Electronic Structure

Electronic structure of C15 Laves phases of ZrV$_2$ were studied with the help of ab initio calculations and the relations between electronic structure and hydrogenation properties were analyzed [7, 8, 49]. It was found that the ZrV$_2$ C15 Laves phase compound, which is cubic at room temperature, undergoes a structural transformation to a non-Laves rhombohedral phase at about 110 K.

Application of ZrV$_2$ as hydrogen storage alloy focused also on the electronic structure and its modification by Ni impurities (Tables 11.4, 11.5) [49]. A number of band calculations were performed since 1980 up to now because of superconducting properties of ZrV and its derivatives [5, 49, 50]. The Ni atom, which replaces V, gives additional 5 electrons per unit cell to the valence band. The valence band for Zr(V$_{0.75}$Ni$_{0.25}$)$_2$ is wider (see Figs. 11.10 and 11.11) than in the case of ZrV$_2$ and the main peak from Ni electrons is located about 2 eV below the Fermi level (E_F, in our case $E_F = 0$). The bottom of the valence band is shifted by about 0.5 eV toward lower energies (from 6.092 to 6.629 eV below E_F). Also the centers of gravity of the s and d electrons located on Zr and V atoms are slightly shifted to lower energies. The charge transfer is from Zr and V atoms to the Ni atoms. A filling of the d-band of Ni atoms is observed. The Fermi level for the doped system is located near the valley of the density of states (DOS) on the increasing slope of the DOS. The value of DOS ($E = E_F$) decreases form 13.873 to 8.882 states/(eV unit cell) for the pure and doped system, respectively. From the analysis of the DOS and their partial wave decomposition around the different atomic sites, we found that the occupied bands in Zr(V$_{1-x}$Ni$_x$)$_2$ compounds are mostly composed of d-type electrons (see Table 11.4). The contributions of d-electrons to the total values of DOS at the Fermi level are 71% and 83.9% for the ZrV$_2$ and Zr$_2$V$_3$Ni system, respectively.

Table 11.4 Number of valence electrons on Zr, V, and Ni atoms for the atomic configuration and for $Zr(V_{1-x}Ni_x)_2$ compounds, where $x = 0$ and 0.25 (in the brackets numbers of d electrons)

Atomic configurations	Zr $5s^2\,4d^2$	V $5s^2\,4d^3$	Ni $5s^2\,4d^8$
ZrV_2	4.4151 (2.7191)	4.7924 (3.5741)	—
$Zr(V_{0.75}Ni_{0.25})_2$	4.2802 (2.5951)	4.7025 (3.4977)	10.3323 (8.5881)

Table 11.5 The values of DOS at the Fermi level for $Zr(V_{1-x}Ni_x)_2$ compounds, where $x = 0$ and 0.25. Total and local values on Zr, V, and Ni atoms and the contributions of d electrons to the values (all values are calculated in states per eV and unit cell)

	Total	Total d-electrons	Zr atom total	Zr atom d-electrons	V atoms total	V atoms d-electrons	Ni atoms total	Ni atoms d-electrons
ZrV_2	13.87	9.91	4.39	2.43	9.48	7.48	—	—
$Zr(V_{0.75}Ni_{0.25})_2$	8.88	7.42	2.02	1.50	5.56	4.90	1.30	1.02

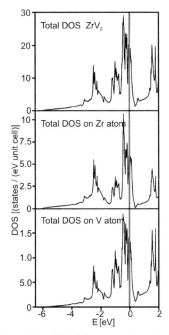

Figure 11.10 The total and local DOS functions for the ZrV_2 system.

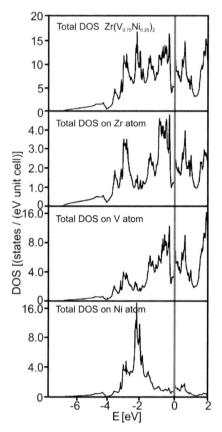

Figure 11.11 The total and local DOS functions for the $Zr(V_{0.75}Ni_{0.25})_2$ system.

11.5 Zr-Based Alloys with MWCNT Addition

For practical applications, hydrogen absorption properties of Zr-based Laves phase alloys need to be further improved. For example, Ni substitution has been examined in many aspects of hydrogen storage alloys. Ni is believed to act as a catalyst and has received considerable attention [5, 51–56]. The partial addition of Ni has been proved to be an efficient way to improve the hydrogenation properties of Zr-based alloys [55, 56].

The work of Jain et al. has shown plateau pressure rises and cell parameters and enthalpy of dissolved hydrogen in $ZrFe_{2-x}Ni_x$-H decreases with the increase of Ni concentration; thus the less hysteresis between hydrogen absorption and desorption can be expected [57].

The use of catalysts has been proved to be an important pathway to enhance surface kinetics for hydrogen storage alloys. Carbon nanotubes (CNTs) have been found to exhibit excellent catalytic effects on the hydrogen sorption properties of hydrogen storage composites [60–63]. Addition of multi-wall carbon nanotubes (MWCNTs) by HEBM offers some reasons for storing hydrogen with high capacity and fast kinetics at relatively low temperatures.

The large surface areas of MWCNTs can assist the dissociation of gaseous hydrogen and finally the small volume of particles after HEBM produces short diffusion paths to the alloys' interiors. In addition, H-atoms can diffuse into the particles along the tubular structure of MWCNTs. It has already been recognized that the combination of MWCNTs with hydrogen storage alloys has been found to lead to an especially great enhancement of hydrogen dissociation and diffusion for hydrogen storage alloys [60, 62].

Recently, the microstructure and hydrogenation properties of a Zr-based Laves phase with the composition of $Zr(V_{0.95}Ni_{0.05})_2$ milled with MWCNTs were studied [63]. The as-milled $Zr(V_{0.95}Ni_{0.05})_2$ alloys with different concentrations of MWCNTs preserve a ZrV_2 phase with C15 type structure. The lattice parameter of C15-type cubic ZrV_2 increases with increasing MWCNT concentration, which partly originates from the dissolution of MWCNTs into the crystal lattice of ZrV_2. It is also found that the addition of MWCNTs increases hydrogenation kinetics in the way of increasing diffusion rate of H atoms caused by MWCNTs with hollow structures. In addition, the width of plateau and the equilibrium pressure of $Zr(V_{0.95}Ni_{0.05})_2$ increase with increasing MWCNT concentration. Meanwhile, the absolute value of enthalpy and entropy decreases as the MWCNT concentration increases from 1 to 10 wt%, indicating that the addition of MWCNTs dramatically decreases the stability of $Zr(V_{0.95}Ni_{0.05})_2$.

References

1. Buschow, K. H. J., Bouten, P. C. P., and Miedema, A. R. (1982). Hydrides formed from intermetallic compounds of two transition metals: A special class of ternary alloys. *Rep. Prog. Phys.* **45:** 937–1039.

2. Sandrock, G., Suda, S., Schlapbach, L. (1992). In Schlapbach, L., ed., *Hydrogen in Intermetallic Compounds II, Topics in Applied Physics*, vol. 67, Chapter 5, Springer-Verlag.
3. Anani, A., Visintin, A., Petrov, K., Srinivasan, S., Reilly, J. J., Johnson, J. R., Schwarz, R. B., and Desch, P. B. (1994). Alloys for hydrogen storage in nickel/hydrogen and nickel/metal hydride batteries. *Power Sources* **47**: 261–275.
4. Sakai, T., Matsuoka, M., and Iwakura, C. (1995). In Gschneidner, K. A., Eyring, L., eds., *Handbook on the Physics and Chemistry of Rare Earths*, vol. 21. Elsevier, Amsterdam, pp. 135–180.
5. Kleperis, J., Wójcik, G., Czerwinski, A., Skowronski, J., Kopczyk, M., and Beltowska-Brzezinska, M. (2001). Electrochemical behavior of metal hydrides. *J. Solid State Electrochem.* **5**: 229–249.
6. Nakano, H., Wakao, S., Shimizu, T., and Morii, K. (1998). Development of C14-type Laves phase alloy for high-rate battery application. *J. Electrochem. Soc. Jpn.* **66**: 734–739.
7. Young, K., Fetcenko, M. A., Koch, J., Morii, K., and Shimizu, T. (2009). Studies of Sn, Co, Al, and Fe additives in C14/C15 Laves alloys for NiMH battery application by orthogonal arrays. *J. Alloys Comp.* **486**: 559–569.
8. Yu, J. Y., Lei, Y. Q., Chen, C. P., Wu, J., and Wang, Q. D. (1995). The electrochemical properties of hydrogen storage Zr-based Laves phase alloys. *J. Alloys Comp.* **231**: 578–581.
9. Joubert, J. M., Sun, D., Latroche, M., and Percheron-Guégan, A. (1997). Electrochemical performances of ZrM_2 (M = V, Cr, Mn, Ni) Laves phases and the relation to microstructures and thermodynamical properties. *J. Alloys Comp.* **253–254**: 564–569.
10. Visintin, A., Peretti, H. A., Ruiz, F., Corso, H. L., and Triaca, W. E. (2007). Effect of additional catalytic phases imposed by sintering on the hydrogen absorption behavior of AB_2 type Zr-based alloys. *J. Alloys Comp.* **428**: 244–251.
11. Young, K., Nei, J., Ouchi, T., and Fetcenko, M. A. (2011). Phase abundances in AB_2 metal hydride alloys and their correlations to various properties. *J. Alloys Comp.* **509**: 2277–2284.
12. Yang, X. G., Zhang, W. K., Lei, Y. Q., and Wang, Q. D. (1999). Electrochemical properties of Zr–V–Ni system hydrogen storage alloys. *J. Electrochem. Soc.* **146**: 1245–1250.
13. Yoshida, M., and Akiba, E. (1995). Hydrogen absorbing-desorbing properties and crystal structure of the Zr-Ti-Ni-Mn-V AB_2 Laves phase alloys. *J. Alloys Comp.* **224**: 121–126.

14. Shu, K., Zhang, S., Lei, Y., Lü, G., and Wang, Q. (2003). Effect of Ti on the structure and electrochemical performance of Zr-based AB_2 alloys for nickel-metal rechargeable batteries. *J. Alloys Comp.* **349**: 237–2341.
15. Takeshita, H. T., Kiyobayashi, T., Tanaka, H., Kuriyama, N., and Haruta, M. (2000). Reversible hydrogen absorption and desorption achieved by irreversible phase transition. *J. Alloys Comp.* **311**: L1–L4.
16. Shi, Z., Chumbley, S, and Laabs, F. C. (2000). Electron diffraction analysis of an AB_2-type Laves phase for hydrogen battery applications. *J. Alloys Comp.* **312**: 41–52.
17. Massalski, T. B. (1996). Structure and stability of alloys. In Cahn, R. W., Hassen, P., ed., *Physical Metallurgy*, Amsterdam: Elsevier, pp. 135–204.
18. Rudy, E. (1969). Compendium of Phase Diagram Data, U.S. Report AFML-TR-65-2, Part V, Air Force Materials Laboratory, Wright Patterson AFB, 8 and 75–76.
19. Williams, J. T. (1955). Vanadium-zirconium alloy system. *Trans. Metall. Soc. AIME*, **203**: 345–350.
20. Kozhanov, V. N., Pushkin, V. B., Romanov, Y. P., Romanova, R. R., and Syutkina, N. N. (1982). Zr-C phase diagram. *Phys. Metals Metallogr.* **53**: 41–46.
21. Bulakh, I. E., Gabovich, A. M., Morozovskii, A. E., Pan, V. M., and Shpigel, S. S. (1983). Zr–V phase diagram. *Sov. Phys. Solid State* **2**: 504–505.
22. Laves, F. (1956). Theory of alloy phases. ASM. Cleveland.
23. Thoma, D. J., and Perepezko, J. H. (1992). An experimental evaluation of the phase relationships and solubilities in the Nb–Cr system. *Mater. Sci. Eng. A* **156**: 97–108.
24. Hong, S., and Fu, C. L. (1999). Phase stability and elastic moduli of Cr_2Nb by first-principles calculations. *Intermetallics* **7**: 5–9.
25. Stein, F., Palm, M., and Sauthoff, G. (2004). Structure and stability of Laves phases. Part I. Critical assessment of factors controlling Laves phase stability. *Intermetallics* **12**: 713–720.
26. Zhu, J. H., Liaw, P. K., and Liu, C. T. (1997). Effect of electron concentration on the phase stability of $NbCr_2$-based Laves phase alloys. *Mater. Sci. Eng. A* **239–240**: 260–264.
27. Berry, R. L., and Raynor, G. V. (1953). The crystal chemistry of the Laves phases. *Acta Crystallogr.* **6**: 178–186.
28. Nevitt, M. V. (1963). Alloy chemistry of transition elements. In Beck, P. A., ed., *Electronic Structure and Alloy Chemistry of the Transition Elements*, New York: Interscience Publishers. pp. 101–178.

29. Zhu, J., Liu, C., Pike, L., and Liaw, P. (1999). A thermodynamic interpretation of the size-ratio limits for Laves phase formation. *Metall. Mater. Trans. A* **30**: 1449–1452.

30. Zhu, J. H., Liu, C. T., Pike, L. M., and Liaw, P. K. (2002). Enthalpies of formation of binary Laves phases. *Intermetallics* **10**: 579–595.

31. Thoma, D. J., and Perepezko, J. H. (1995). A geometric analysis of solubility ranges in Laves phases. *J. Alloys Comp.* **224**: 330–341.

32. Jurczyk, M., Rajewski, W., Majchrzycki, W., and Wojcik, G. (1998). Synthesis and electrochemical properties of high energy ball milled Laves phase $(Zr,Ti)(V,Mn,Cr)_2$ alloys with nickel powder. *J. Alloys Comp.* **274**: 299–302.

33. Jurczyk, M., Rajewski, W., Wojcik, G., and Majchrzycki, W. (1999). Metal hydride electrodes prepared by mechanical alloying of ZrV_2-type materials. *J. Alloys Comp.* **285**: 250–254.

34. Jurczyk, M., and Nowak, M. (2008). Nanomaterials for hydrogen storage synthesized by mechanical alloying. In: Ali, E., ed., *Nanostructured Materials in Electrochemistry* (Wiley), Chapter 9.

35. Majchrzycki, W., and Jurczyk, M. (2001) Electrode characteristics of nanocrystalline $(Zr,Ti)(V,Cr,Ni)_{2.41}$ compound. *J. Power Sources* **93**: 77–81.

36. Majchrzycki, W. (2002). The influence of microstructure on electrochemical properties of some alloys in Ni-MH applications. PhD thesis, Poznan University of Technology.

37. Au, M., Pourarian, F., Simizu, S., Sankar, S. G., and Zhang, L. (1995). Electrochemical properties of $TiMn_2$-type alloys ball-milled with nickel powder. *J. Alloys Comp.* **223**: 1–5.

38. Jurczyk, M., Smardz, L., and Szajek, A. (2004). Nanocrystalline materials for NiMH batteries, *Mater. Sci. Eng. B* **108**: 67–75.

39. Jurczyk, M. (2004).The progress of nanocrystalline hydride electrode materials. *Bull. Pol. Acc., Tech.* **52**: 67–77.

40. Jurczyk, M., Smardz, L., Makowiecka, M., Jankowska, E., Smardz, K. (2004). The synthesis and properties of nanocrystalline electrode materials by mechanical alloying. *J. Phys. Chem. Sol.* **65**, 545–548.

41. Zaluski, L., Zaluska, A., and Ström-Olsen, J. O. (1997). Nanocrystalline metal hydrides. *J. Alloys Comp.* **253–254**: 70–79.

42. Zaluska, A., Zaluski, L., and Strom-Olsen, J. O. (1999). Synergy of hydrogen sorption in ball-milled hydrides of Mg and Mg_2Ni. *J. Alloys Comp.* **289**: 197–206.

43. Jurczyk, M., Jankowska, E., Nowak, M., Jakubowicz, J. (2002). Nanocrystalline titanium type metal hydrides prepared by mechanical alloying. *J. Alloys Comp.* **336**: 265–269.
44. Selvam, P., Viswanathan, B. (1990). Surface properties and their consequences on the hydrogen sorption characteristics of certain materials. *J. Less Common Metals* **163**: 89–108.
45. Young, K., Ouchi, T., Huang, B., Chao, B., Fetcenko, M. A., Bendersky, L. A., et al. (2010). The correlation of C14/C15 phase abundance and electrochemical properties in the AB2 alloys. *J. Alloys Comp.* **506**: 84184–84188.
46. Kronberger, H. (1997). Nanocrystalline hydrogen storage alloys for rechargeable batteries. *J. Alloys Comp.* **253–254**: 87–89.
47. Jurczyk, M., Rajewski, W. (2000). Nanostructured hydrogen storage materials synthesized by mechanical alloying. In Rühle, M., and Gleiter, H., ed., *Interface Controlled Materials*, vol. **9,** Willey-VCH Verlag Gmbh.
48. Jurczyk, M., Majchrzycki, W., Nowak, M., and Jankowska, E. (2001). Electrochemical properties of nanocrystalline (Zr,La)(V,Ni)2.25 alloy. *J. Alloys Comp.* **322**: 233–237.
49. Szajek, A., Jurczyk, M., Rajewski, W. (2000). The electronic and electrochemical properties of the ZrV_2 and $Zr(V_{0.75}Ni_{0.25})_2$ systems. *J. Alloys Comp.* **302**: 299–303.
50. Jezierski, A., Jegierski, A., and Coldea, M. (1998). Electronic structure of $Hf_{1-x}Zr_xV_2$ alloys. *Phys. Stat. Sol. B* **208**: 87–90.
51. Jain, I. P. (2009). Hydrogen the fuel for 21st century. *Int. J. Hydrogen Energy* **34**: 7368–7378.
52. Lv, Y. J., Zhang, B., and Wu, Y. (2015). Effect of Ni content on microstructural evolution and hydrogen storage properties of Mg-xNi-3La (x = 5, 10, 15, 20 at%) alloys. *J. Alloy Comp.* **641**: 176–180.
53. Huang, J. M., Ouyang, L. Z., Wen, Y. J., Wang, H., Liu, J. W., Chen, Z. L., et al. (2014). Improved hydrolysis properties of Mg_3RE hydrides alloyed with Ni. *Int. J. Hydrogen Energy* **39**: 6813–6818.
54. Young, M., Chang, S., Young, K., Nei, J. (2013). Hydrogen storage properties of $ZrV_xNi_{3.5-x}$ (x = 0.0–0.9) metal hydride alloys. *J. Alloys Comp.* **580**: S171—S174.
55. Kumar, V., Pukazhselvan, D., Tyagi, A. K., and Singh, S. K. (2013). Effect of Ni concentration on the structural and hydrogen storage characteristics of Zr-Mn based Laves phase system. *Mater. Renew. Sustain. Energy* **2**: 1–8.

56. Jat, R. A., Parida, S. C., Agarwal, R., and Kulkarni, S. G. (2013). Effect of Ni content on the hydrogen storage behavior of ZrCo$_{1-x}$Ni$_x$ alloys. *Int. J. Hydrogen Energy* **38**: 1490–1500.

57. Jain, A., Jain, R. K., Agarwal, S., Ganesan, V., Lalla, N. P., Phase, D. M., et al. (2007). Synthesis, characterization and hydrogenation of alloys. *Int. J. Hydrogen Energy* **32**: 3965–3971.

58. Hou, X. J., Hu, R., Zhang, T. B., Kou, H. C., and Li, J. S. (2013). Hydrogenation behavior of high-energy ball milled amorphous Mg$_2$Ni catalyzed by multi-walled carbon nanotubes. *Int. J. Hydrogen Energy* **38**: 16168–16176.

59. Shaijumon, M. M., Rajalakshmi, N., Ryu, H., and Ramaprabhu, S. (2005). Synthesis of multi-walled carbon nanotubes in high yield using Mm based AB$_2$ alloy hydride catalysts and the effect of purification on their hydrogen adsorption properties. *Nanotechnology* **16**: 518–524.

60. Balcerzak, M., Jakubowicz, J., Kachlicki, T., and Jurczyk, M. (2015). Effect of multi-walled carbon nanotubes and palladium addition on the microstructural and electrochemical properties of the nanocrystalline Ti$_2$Ni alloy. *Int. J. Hydrogen Energy* **40**: 3288–3299.

61. Balcerzak, M., Jakubowicz, J., Kachlicki, T., and Jurczyk, M. (2015). Hydrogenation properties of nanostructured Ti$_2$Ni-based alloys and nanocomposites. *J. Power Sources* **280**: 435–445.

62. Ranjbar, A., Ismail, M., Guo, Z. P., Yu, X. B., and Liu, H. K. (2010). Effects of CNTs on the hydrogen storage properties of MgH$_2$ and MgH$_2$-BCC composite. *Int. J. Hydrogen Energy* **35**: 7821–7826.

63. Wu, T. D., Xue, X. G., Zhang, T. B., Hu, R., Kou, H. C., and Jinshan Li, J. H. (2016). Effect of MWCNTs on hydrogen storage properties of a Zr-based Laves phase alloy. *Int. J. Hydrogen Energy*, available online 1 February 2016; doi:10.1016/j.ijhydene.2015.12.114.

Chapter 12

LaNi$_5$-Based Hydrogen Storage Alloys

Marek Nowak and Mieczyslaw Jurczyk

*Institute of Materials Science and Engineering,
Poznan University of Technology, Poznan, Poland*

marek.nowak@put.poznan.pl, mieczysław.jurczyk@put.poznan.pl

12.1 Phase Diagram and Structure

The La–Ni phase diagram is shown in Fig. 12.1. The recent version of the diagram shows that eight intermetallic phases, La$_3$Ni, La$_7$Ni$_3$, LaNi, La$_2$Ni$_3$, La$_7$Ni$_{16}$, LaNi$_3$, La$_2$Ni$_7$, and LaNi$_5$, exist in the equilibrium state (Table 12.1) [1]. No stable intermetallic phase has been reported to exist between La$_2$Ni$_7$ and LaNi$_5$. LaNi$_5$ has the CaCu$_5$-type hexagonal structure with a = 5.017 Å and c = 3.987 Å, while La$_2$Ni$_7$ has the Ce$_2$Ni$_7$-type hexagonal structure with a = 5.058 Å and c = 24.71 Å [2–4]. The unit cell of La$_2$Ni$_7$ is considered to be made of two block layers, each consisting of one unit layer of the LaNi$_5$-type and two unit layers of the LaNi$_2$ (Laves)-type [3–6] (see Fig. 12.2).

Recently, microstructure and hydrogen absorption/desorption properties of La–Ni alloys have been investigated as a function of alloy composition in the range of La-77.8–83.2 at% Ni, which corresponds to compositions between two intermetallic phases,

---

Handbook of Nanomaterials for Hydrogen Storage
Edited by Mieczyslaw Jurczyk
Copyright © 2018 Pan Stanford Publishing Pte. Ltd.
ISBN 978-981-4745-66-6 (Hardcover), 978-1-315-36444-5 (eBook)
www.panstanford.com

La$_2$Ni$_7$ and LaNi$_5$ [6]. The intermetallic phase, La$_5$Ni$_{19}$, of Ce$_5$Co$_{19}$-type is found for the first time to exist as an equilibrium phase at a composition between La$_2$Ni$_7$ and LaNi$_5$. This phase is stable at high temperatures around 1000°C but decomposes into La$_2$Ni$_7$ and LaNi$_5$ below 900°C. Hydrogen absorption/desorption properties described in terms of pressure–composition isotherms decline with decreasing Ni content (i.e., with increasing volume fraction of intermetallic phases other than LaNi$_5$). In particular, the plateau at the equilibrium pressure corresponding to the hydrogen absorption in the LaNi$_5$ phase is narrowed with decreasing Ni content and additional plateaus with higher equilibrium pressures come into existence. The degradation becomes more pronounced in the presence of La$_2$Ni$_7$ than La$_5$Ni$_{19}$. This can be understood in terms of the ratio of the number of LaNi$_2$ (Laves) unit layers to that of LaNi$_5$ unit layers in the unit cell of the two intermetallic phases [6].

LaNi$_5$ is the most studied compound as a hydrogen storage material. The LaNi$_5$ alloy absorbs about 1.0 H/LaNi$_5$ (1.5 wt%) and is easy to activate (5 bar at T = 100°C after a few cycles, to achieve the maximum hydrogen capacity). The PCT diagram shows a flat plateau, low hysteresis, but unfortunately the hydrogen capacity is degraded after a few cycles. In the LaNi$_5$-related intermetallic compounds, the thermodynamic and electrochemical properties were found to be strongly dependent on the nature of the substituted element, its content, and its microstructure.

Table 12.1 La–Ni crystal structures data

Phase	Composition [at%] Ni	Prototype
La$_3$Ni	25.0	Fe$_3$C
La$_7$Ni$_3$	30.0	Fe$_3$Th$_7$
LaNi	50.0	CrB
La$_2$Ni$_3$	60.2	—
La$_7$Ni$_{16}$	69.6	—
LaNi$_3$	75.0	PuNi$_3$
βLa$_2$Ni$_7$	77.8	Ga$_2$Ni$_7$
αLa$_2$Ni$_7$	77.8	Ce$_2$Ni$_7$
LaNi$_5$	83.3	CaCu$_5$

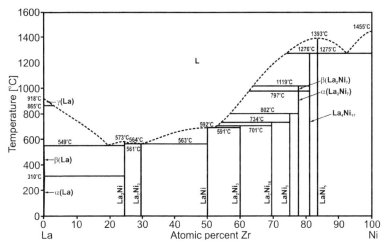

Figure 12.1 La–Ni phase diagram [1].

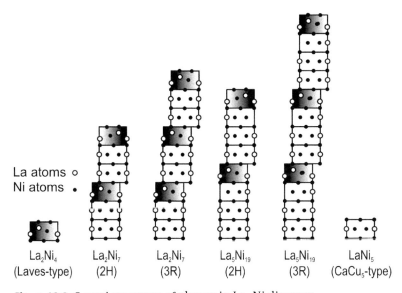

Figure 12.2 Crystal structures of phases in La–Ni diagram.

12.2 LaNi$_5$-Type Compounds

The exceptional hydriding properties of LaNi$_5$ were first reported by Van Vucht et al. [7], who found at room temperature that LaNi$_5$ rapidly and reversibly absorbed more than six atoms of hydrogen per formula unit. Since then, this alloy system has become the most widely used intermetallic alloy in all metal hydride applications. The LaNi$_5$ alloy crystallizes in the hexagonal CaCu$_5$-type structure. The merit of these compounds is that they exhibit low hysteresis, are tolerant to gaseous impurities, and are easily hydrogenated in the initial cycle after manufacture. The properties of hydrogen host materials can be modified substantially by alloying, to obtain the desired storage characteristics, e.g., proper capacity at a favorable hydrogen pressure. Until now, many multi-component La$_{1-x}$RE$_x$Ni$_{5-x}$M$_x$ systems (RE = mischmetal, rare earth; M = Al, Mn, Si, Zn, Cr, Fe, Cu, Co, Sn and Ge) have been studied with the double goal of both changing the stability of their hydrides and understanding which parameters play a role in the stability and hydrogen content of their hydrides [8–46].

Although the La$_{1-x}$RE$_x$Ni$_5$ solid solutions, where RE represents a rare earth, exist over an entire range of x values, the LaNi$_{5-x}$M$_x$ solid solutions do not necessarily exist for all values of x. In the latter case, the maximum value of x varies according to the substitutional element, M, and is 0.6 for silicon, 1.2 for iron, 1.3 for aluminum, 2.2 for manganese, and 5 for copper and cobalt. This maximum value depends on the heat treatment of the sample which has to be optimized for each family of alloys. For example, this maximum can be increased by lowering the annealing temperature in the case of LaNi$_{5-x}$Mn$_x$ or by increasing it in the case of LaNi$_{5-x}$Fe$_x$. However, it seems that the atomic size and the electronic structure of the M atom are the main factors limiting the value of x.

All of the La$_{1-x}$RE$_x$Ni$_5$ and LaNi$_{5-x}$M$_x$ solid solutions crystallize in the hexagonal CaCu$_5$ structure with unit cell dimensions which vary with RE and M. For a given solid solution, the unit cell volume varies linearly with x, and specifically it decreases in the La$_{1-x}$RE$_x$Ni$_5$ solid solutions, where as it increases in the LaNi$_{5-x}$M$_x$ solid solutions.

It was found that the substitution of Ni in LaNi$_5$ by small amounts of M-type metals altered the hydrogen storage capacity, the stability of the hydride phase, and the corrosion resistance. Generally, in the transition metal sublattice of LaNi$_5$-type compounds, substitution by Mn, Al, and Co has been found to offer the best compromise between high hydrogen capacity and good resistance to corrosion [38–42].

In the $LaNi_{5-x}M_xH_y$ hydrides, the hydrogen capacity, y, may be strongly affected by the type and concentration of the substituted M metal. When aluminum is substituted for nickel, y decreases from 6.2 in $LaNi_{4.9}Al_{0.1}$ to 4.8 in $LaNi_4Al$ [36]. On the other hand, when manganese is substituted for nickel, y is nearly constant at 6 over the range of x values between 0 and 2. Additionally, the hydrogen content is less affected by the substitution of lanthanum. A nearly constant y value of approximately 6.2 at 40°C is found for the series of $La_{0.8}RE_{0.2}Ni_5$ compounds, where RE is Y, Nd, Gd, and Er [32].

In these substituted hydrides, the unit cell volume expansion is smaller than in $LaNi_5$ and is 20% in $LaNi_4MnH_6$ and 14.4% in $LaNi_4AlH_{4.5}$. These expansions correspond to 3.3 and 3.0 $Å^3$ per hydrogen atom, compared with $3.8Å^3$ per hydrogen atom observed for $LaNi_5H_{6.5}$. The formation of these hydrides leads to internal lattice microstrains which do not relax during the hydrogen desorption process.

The investigation of the $La_{1-x}Ce_xNi_5$ (0.3 < x < 0.0) system shows that with increasing cerium content: (1) there is an important decrease of the cell volume, (2) the hydrogenation kinetics were found to be lower when (3) both plateau pressure and hysteresis increase [16]. After CO surface treatment, the retention time of H_2 under normal conditions was extended. Increasing Ce content in CO-treated hydrides leads to a decrease in the retention time: 820 ks for complete dehydrogenation of CO-treated $LaNi_5$ hydride compared to 110 ks for CO-treated $La_{0.9}Ce_{0.1}Ni_5$ hydride.

12.3 LaNi₅ Phase Synthesized by Mechanical Alloying

Mechanical alloying is a suitable method for the synthesis of nanocrystalline AB_5 alloys [42, 46]. Because of the large difference in melting point between La and Ni, mechanical alloying could be a good method for the synthesis of these compounds. Compared to the microcrystalline alloy prepared by arc- or induction melting, mechanically alloyed $LaNi_5$ is easier to activate and has faster first hydrogenation kinetics.

During the MA process, originally sharp diffraction lines of La and Ni gradually become broader and their intensity decreases with milling time (Fig. 12.3). The powder mixture milled for more than 30 h transforms completely to the amorphous phase [42, 46]. Formation of the nanocrystalline alloy was achieved by annealing of the amorphous material in high-purity argon atmosphere at 700°C for 0.5 h (Fig. 12.3c). According to AFM (Fig. 12.4) and TEM studies, the average size of annealed La–Ni powders was of order of 25 nm (Fig. 12.5).

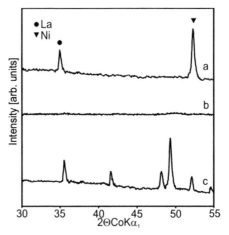

Figure 12.3 X-ray diffraction spectra of La and Ni powders during MA process: (a) 0 h, (b) after 30 h and (c) after heat treatment at 700°C/0.5 h.

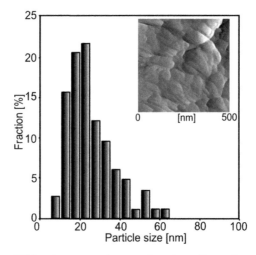

Figure 12.4 AFM picture and particle size dimension distribution histograms of LaNi$_5$ powdered sample (30 h MA, heat treated at 700°C for 0.5 h).

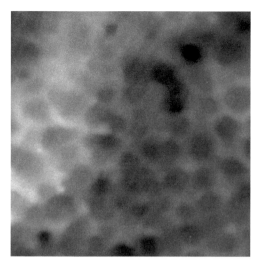

Figure 12.5 TEM micrograph of the mechanically alloyed for 30 h and annealed at 700°C for 0.5 h LaNi$_5$ sample (× 64 000).

The LaNi$_5$ alloy can absorb up to 5.5 H/f.u. at room temperature. X-ray diffraction spectra of LaNi$_4$Al$_1$ alloy before (a) and after hydrogenation are presented in Fig. 12.6.

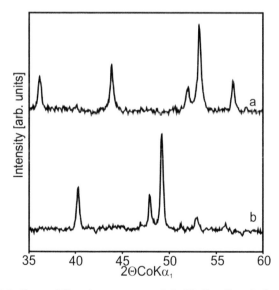

Figure 12.6 X-ray diffraction spectra of LaNi$_4$Al$_1$ alloy before (a) and after hydrogenation (b).

12.4 XPS and AES Studies

Figure 12.7 shows XPS spectrum of the freshly prepared nanocrystalline 50 nm-LaNi$_5$ thin film. The sample was deposited after the outgassing procedure of the glass substrate with holder at 430°C for 3 h. For a comparison, we show results for the nanocrystalline LaNi$_5$ bulk alloy. The spectrum of the thin film was measured in situ, immediately after deposition. The bulk sample was prepared ex situ by MA followed by annealing in a high-purity argon atmosphere [47, 48]. The bulk nanocrystalline LaNi$_5$ was measured immediately after heating in UHV conditions followed by the removal of a native oxide and possible impurities layer using an ion gun etching system. The XPS intensity of the MA nanocrystalline alloy is considerably reduced compared to that measured for a thin film. This is due to not perfectly clean surface and significantly greater roughness parameter of the bulk alloy. The oxygen concentration on the surface is about 0.1 and 2 at% for nanocrystalline LaNi$_5$ thin film and MA bulk alloy, respectively. The XPS measurements after the removal of the 50 nm top LaNi$_5$ layer revealed the same oxygen concentration

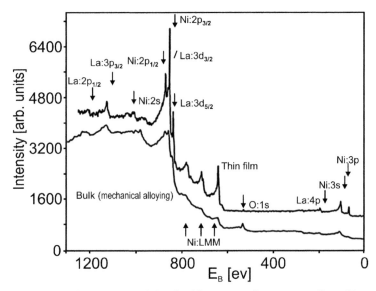

Figure 12.7 XPS spectrum of the freshly prepared nanocrystalline 50 nm LaNi$_5$ thin film. For a comparison, we show results for the MA nanocrystalline LaNi$_5$ bulk alloy.

in the MA sample while practically no trace of oxygen atoms in the thin film volume. The average roughness parameter determined by scanning tunneling microscopy (STM) was about 1 and 100 nm for nanocrystalline LaNi$_5$ thin film and MA bulk alloy, respectively.

The cleanliness of the surface of nanocrystalline LaNi$_5$-type alloys was studied by X-ray photoelectron spectroscopy (XPS) and Auger electron spectroscopy (AES) [47, 49]. Figure 12.8 shows the element-specific Auger intensities of the nanocrystalline LaNi$_{4.2}$Al$_{0.8}$ sample as a function of the sputtering time, converted to depth. Similar to the results obtained for the microcrystalline sample, there is relatively high concentration of carbon and oxygen immediately on the surface. The carbon concentration strongly decreases toward the interior of the sample. We have found that at the oxide–metal interface, only iron impurities and lanthanum atoms are present. As the escape-depth of the Auger electrons from nickel and aluminum atoms is about 2 nm, we conclude that these elements are practically not present on the metallic surface. In other words, lanthanum atoms and iron impurities strongly segregate to the surface and form an oxide layer under atmospheric conditions. The lower lying Ni atoms form a metallic subsurface layer and are responsible for the observed high hydrogenation rate in accordance with earlier findings. The above segregation process is stronger compared to that observed for the microcrystalline sample. The presence of the significant amount of iron atoms in the surface layer of the nanocrystalline LaNi$_{4.2}$Al$_{0.8}$ alloy could be explained by Fe impurities trapped in the MA powders from the erosion of the milling media. The amount of the Fe impurities considerably decreases in the subsurface layer of the sample. From the peak-to-peak amplitude, we have estimated a maximum atomic concentration of oxygen as ~2 at% in the interior of the sample. Similar to the results reported for the microcrystalline LaNi$_{4.2}$Al$_{0.8}$ alloy, the concentration of carbon impurities inside the sample was below 0.5 at%.

In Fig. 12.9 we show the XPS valence bands for nanocrystalline (bold solid line) and microcrystalline (thin solid line) LaNi$_5$ films. The shape of the valence band measured for the microcrystalline LaNi$_5$ thin film is practically the same compared to that reported earlier for the single crystalline sample. On the other hand, the

XPS valence band of the nanocrystalline LaNi$_5$ thin film (bold solid line) is considerably broader compared to that measured for the microcrystalline sample. This is probably due to a strong deformation of the nanocrystals. Normally the interior of the nanocrystal is constrained and the distances between atoms located at the grain boundaries are expanded. Such a modification of the electronic structure of the nanocrystalline LaNi$_5$-type alloy compared to that of microcrystalline sample could significantly influence its hydrogenation properties [47].

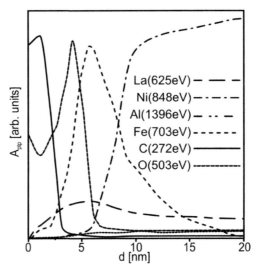

Figure 12.8 AES spectrum of nanocrystalline LaNi$_{4.2}$Al$_{0.8}$ alloy vs. sputtering time, as converted to depth. The sample surface is located on the left-hand side.

Similar broadening of the valence band can be also observed for the MA nanocrystalline LaNi$_{4.2}$Al$_{0.8}$ bulk alloy (bold broken line in Fig. 12.9) compared to that measured for the microcrystalline bulk sample (thin broken line in Fig. 12.9).

Results on XRF measurements revealed the assumed bulk chemical composition of the microcrystalline and nanocrystalline LaNi$_5$-type thin films and bulk materials. However, in [50], we have reported a strong segregation of the La atoms to the surfaces in MA (bulk) nanocrystalline LaNi$_{4.2}$Al$_{0.8}$ alloys. Oxygen-induced surface segregation in LaNi$_5$ alloy was also very well-documented effect.

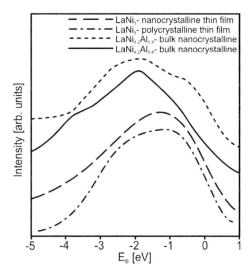

Figure 12.9 XPS valence band spectra for nanocrystalline (bold solid line) and polycrystalline (thin solid line) LaNi$_5$ thin films. For a comparison we show results for the nanocrystalline and polycrystalline LaNi$_{4.2}$Al$_{0.8}$ bulk alloys represented by bold and thin broken lines, respectively.

In Fig. 12.10a, we show the element-specific Auger intensities of the bulk microcrystalline LaNi$_5$ sample as a function of the sputtering time, converted to depth. As can be seen, there is relatively high concentration of carbon and oxygen immediately on the surface, which could be due to carbonates or adsorbed atmospheric CO$_2$. The carbon concentration strongly decreases toward the interior of the sample. At the metal interface itself, only oxygen is present, making it very likely that only an oxide layer is formed, and no other compounds, the latter growing apparently with a smaller probability. We have found that at the oxide–metal interface, mainly lanthanum and nickel atoms are present. Taking into account that the escape depth of the Auger electron from nickel atoms is about 2 nm, the concentration of these elements on the metallic surface is significantly lower compared to the average bulk composition. Therefore, the lanthanum atoms, which segregate to the surface, form a La-based oxide layer under atmospheric conditions. The oxidation process is depth limited such that an oxide-covering layer with a well-defined thickness is formed by which the lower lying metal is prevented from further oxidation. In this way, one can obtain a self-stabilized oxide–metal structure. From the peak-to-peak amplitude, we have calculated a maximum atomic concentration of oxygen inside the sample as 2 at%. The concentration of carbon impurities inside the sample was below 0.5 at%.

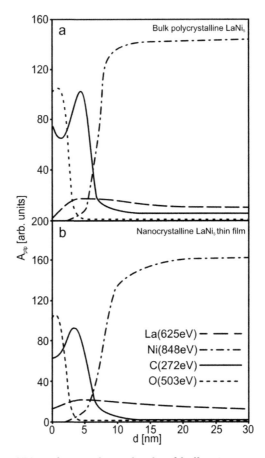

Figure 12.10 AES peak-to-peak amplitude of bulk microcrystalline LaNi$_5$ (a) and nanocrystalline LaNi$_5$ thin film (b) versus sputtering time, as converted to depth. The sample surface is located on the left-hand side. The bulk and thin film samples were oxidized under environmental condition before the AES measurements. We have not heated the samples in UHV to remove adsorbed surface impurities.

As described above, the in situ XPS studies showed no remarkable segregation of La atoms to the surface in nanocrystalline and microcrystalline LaNi$_5$ thin films. On the other hand, AES studies with depth profiling show the La segregation effect after 24 h thin film exposition in the environmental (contact with atmosphere) condition. Figure 12.10b shows the element-specific Auger intensities of the nanocrystalline $LaNi_5$ thin film (after 24 h exposition) as a function of the sputtering time, converted to depth. Similar to the results obtained for the microcrystalline bulk sample, there is relatively high concentration of carbon and oxygen immediately on the surface. The carbon concentration strongly decreases toward the interior of the sample. We have found that at the oxide–metal interface only lanthanum atoms are present. As the escape depth of the Auger electrons from nickel atoms is about 2 nm, we conclude that these elements are practically not present on the metallic surface. In other words, lanthanum atoms segregate to the surface and form an oxide layer under atmospheric conditions. The above segregation process is weaker compared to that observed for the bulk microcrystalline sample. This is probably due to significantly lower roughness of the nanocrystalline thin film surface compared to that observed in bulk microcrystalline sample.

12.5 Thermodynamical Properties

The high gravimetric and volumetric storage capacities of $LaNi_5$-type metal hydrides make them ideal hydrogen carrier. In our work, the thermodynamic and kinetic improvements on the hydrogen release properties of nanostructured metal hydrides are investigated (Figs 12.11 and 12.12).

Mechanical alloying drastically reduces the hydrogen storage capacity of $LaNi_5$ but it can be recovered by annealing (Fig. 12.11). Generally, when MA process is performed under an inert atmosphere (high-purity Ag) hydrogen can be readily absorbed. Compared to the unmilled (microcrystalline) alloy, mechanically alloyed $LaNi_{4.2}Al_{0.8}$ is easier to activate and has faster first hydrogenation kinetics. The plateau pressure increases for nanocrystalline materials with reduction in hydrogen capacity and a sloping plateau. Fujii et al. concluded that no improvement in hydrogen sorption properties is expected from nanocrystalline $LaNi_5$ because of the formation of a too stable hydride in the grain boundaries [13].

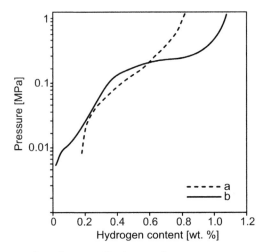

Figure 12.11 p–c–T isotherms of hydrogen desorption at room temperature for LaNi$_{4.2}$Al$_{0.8}$ alloy: (a) an amorphous alloy, (b) nanostructured material.

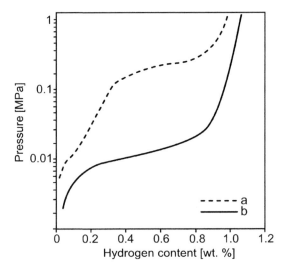

Figure 12.12 p–c–T isotherms at room temperature for LaNi$_{4.2}$Al$_{0.8}$ alloy: (a) nanostructured material, (b) microcrystalline material.

12.6 Electrochemical Properties

The discharge capacity of an electrode prepared by application of nanocrystalline LaNi$_5$ alloy powder is low (Fig. 12.13) [50–52].

It was found that the substitution of Ni by Al or Mn in La(Ni,M)$_5$ alloy leads to an increase in discharge capacity. The LaNi$_4$Mn electrode, mechanically alloyed and annealed, displayed the maximum capacity at the first cycle, but discharge capacity degraded strongly with cycling. On the other hand, alloying elements such as Al, Mn, and Co substituting nickel greatly improved the cycle life of LaNi$_5$-type material.

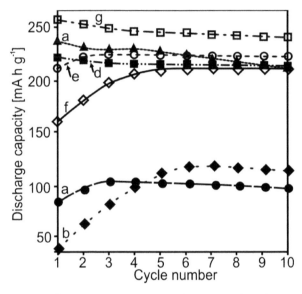

Figure 12.13 Discharge capacities as a function of cycle number of LaNi$_5$-type negative electrodes made from nanocrystalline powders prepared by MA followed by annealing: (a) LaNi$_5$, (b) LaNi$_4$Co, (c) LaNi$_4$Mn, (d) LaNi$_4$Al, (e) LaNi$_{3.75}$CoMn$_{0.25}$, (f) LaNi$_{3.75}$CoAl$_{0.25}$, (g) LaNi$_{3.75}$Mn$_{0.75}$Al$_{0.25}$Co$_{0.25}$ (solution, 6 M KOH; T = 20°C). The charge conditions were 40 mA g^{-1}. The cut-off potential vs. Hg/HgO/6 M KOH was −0.7 V.

With the increase of cobalt content in LaNi$_{4-x}$Mn$_{0.75}$Al$_{0.25}$Co$_x$, the material shows an increase in discharge capacity which passes through a wide maximum for x = 0.25 [46, 53, 54]. In nanocrystalline LaNi$_{3.75}$Mn$_{0.75}$Al$_{0.25}$Co$_{0.25}$ discharge capacities up to 258 mA h g^{-1} (at 40 mA g^{-1} discharge current) was measured, which compares with results reported by Iwakura et al. for microcrystalline MmNi$_{4-x}$Mn$_{0.75}$Al$_{0.25}$Co$_x$ alloys (Mm = mischmetal) [38].

12.7 Electronic Structure

Application of AB$_5$-type alloys as electrode materials focused also on the electronic structure of LaNi$_5$ and its modification by mainly Al and Co atoms [55]. The band structures for LaNi$_5$-type electrodes were calculated using the TB LMTO-ASA method [56, 57] by Szajek et al. [55], with spin–orbit interactions taken into account in the form proposed by Min and Jang [58]. The Perdew–Wang [59] potential with non-local corrections was used. The standard [60] combined corrections terms were included to compensate for errors due to the ASA. The band structures were calculated for experimental values of the lattice constants. In the ASA, the volume of a given unit cell is filled in by Wigner–Seitz spheres having the same total volume:

$$\sum_{A=1}^{N_A} \frac{4}{3}\pi S_A^3 = NA \cdot \frac{4}{3}\pi S_{av}^3 = V,$$

where A is the index of an atom in the unit cell, N_A is number of atoms in the cell, S_A is the Wigner–Seitz radius of the A-type atom, V is volume of the unit cell. In our case the unit cell contains six atoms ($N_A = 6$, one formula unit).

The calculations were performed for all possible positions of impurities in the unit cell. In the case of LaNi$_4$Al the single Al atoms are located in 3g and 2c positions one after the other. The overlap volumes of the muffin-tin spheres for all structures are below 8.5%. The input electronic configurations were taken as: core + 6s^{2}5d^1 for La, core + 4s^{2}3d^7 for Co, core + 3s^{2}3p^1 for Al, and core + 4s^{2}3d^8 for Ni atom. Computations were done for 484, 459, 1224, and 1170 k-points in the irreducible wedge (1/24, 1/12, 1/4 and 1/4) of the first Brillouin zone. It depended on the localization of the impurities changing the symmetry of the system. For integration over the Brillouin zone, the tetrahedron method was used [61]. The iterations were repeated until the energy eigenvalues of the consecutive iteration steps were the same within an error of 0.01 mRy.

The TB-LMTO ASA method allows differentiation of the ions in the cell, but the effect of disorder due to Al and Co substitution is neglected. In the case of LaNi$_5$ system (Haucke-type), two types of Ni sites must be distinguished: the twofold degenerate basal sites (2c) located in the basal plane together with La atoms and the threefold degenerate non-basal sites (3g) located on the plane with z = 0.5 (see Fig. 12.14). In accordance with observation by Gurevitz et al. [62], it is assumed that the La sites do not accommodate Ni, Al, and Co atoms. Our total energy calculations showed that in both cases, LaNi$_4$Al and LaNi$_3$AlCo, the impurity atoms prefer the 3g positions in agreement with the experimental data [63]. For the LaNi$_4$Al system, the difference between the cases when Al is located in 3g and 2c position is about 23.3 mRy (1 mRy ≈ 157 K). Location of Al and Co atoms for the LaNi$_3$AlCo system on the position other than 3g increases the total energy at least by about 10 mRy. The most unstable situation is for the case when impurities are located in 2c positions and then the total energy is increased by about 25 mRy.

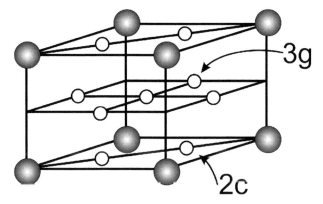

Figurer 12.14 The CaCu$_5$-type unit cell. Small circles describe the Ni(Co,Al) atoms occupying 2c and 3g positions. The La atoms occupy 1a positions described by large circles.

The total densities of states (DOS) and the local contributions of particular atoms are presented in Figs. 12.15 and 12.16 for LaNi$_5$ and LaNi$_3$AlCo, respectively. Presented on Fig. 12.15 results concern the situations when Al and Co atoms or located in 3g positions, e.g., for the most stable situation when the total energies have the minimum values.

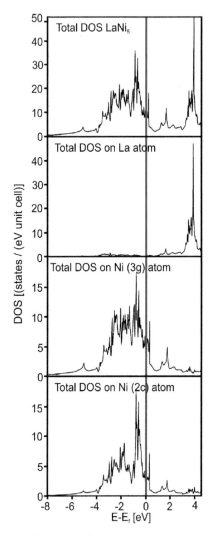

Figure 12.15 Total and local DOS for LaNi$_5$ system.

The width of the valence band of the LaNi$_5$ system is about 7.5 eV. The valence bands for LaNi$_3$CoAl are wider, by about 0.9 eV. It is caused by the Al atom, its s and p electrons are located near the bottom of the valence bands. The values of DOS are lower because of reduced number of valence electrons. Al and Co atoms supply 3 and 9 electrons to the valence band, respectively, whereas Ni atoms 10 electrons, each. The impurities cause charge transfer between the ions. The largest transfer is from the Al atom: about 0.5 electron per atom. For Ni(2c) atoms, the values of charge transfer is much smaller, about 0.07-0.06 electrons per atom. The Ni(3g) atoms are acceptors of electrons, similarly for Co atoms. The transferred electrons are caught mainly by La atoms, which fill in the d bands.

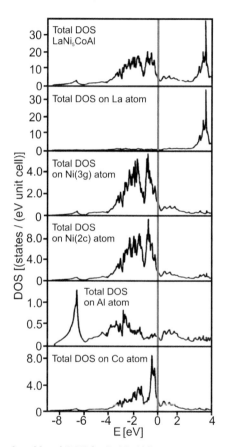

Figure 12.16 Total and local DOS for LaNi$_3$AlCo system.

The Al and Co atoms cause the reconstruction of the starting LaNi$_5$ band structure and the change of the value of DOS at the Fermi level. DOS ($E = E_F$) for LaNi$_5$ is equal to about 10.29 states/(eV f.u.), with about 90% of d electrons contribution. The La contribution is relatively small and for Ni atoms is equal to 2.09 and 1.49 states/(eV atom.) for 3g and 2c sites, respectively.

Sluiter et al. [64] performed band structure calculations to establish the phase stability of La(Ni$_{1-x}$Co$_x$)$_5$ systems in the whole range of concentrations.

The calculations were performed for the ordered LaNi$_5$-LaCo$_5$ compounds. Comparing the total energies associated with replacing a basal (2c) and non-basal sites in LaNi$_5$. Ferromagnetism increased with Co concentration, in accordance with the experimental observation [65]. Whereas LaCo$_5$ is strongly ferromagnetic with a moment of about 1.2 μ_B per Co atom, LaNi$_5$ is non-magnetic. Their calculations showed that even single Co atom surrounded by Ni atoms has a moment equal to 0.435 μ_B, and the moments induced on Ni atoms are below 0.01 μ_B per atom. In our case, for LaNi$_3$AlCo the spin-polarized calculations gave the magnetic moment below 0.001 μ_B/f.u.

12.8 Composite LaNi$_5$-Type Materials

Additionally, the LaNi$_5$/A and Mg$_2$Ni/A (A = graphite, copper, or palladium) nanocomposites were synthesized [66]. The influence of the microstructure on the structural and electrochemical properties of synthesized materials was studied. These measurements may provide useful indirect information about the influence of the surface chemical composition, crystal structure, and grain sizes on the hydrogenation properties of the studied materials (Fig. 12.17).

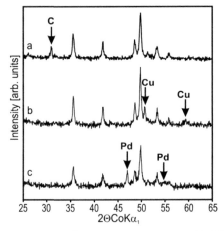

Figure 12.17 XRD spectra of nanocomposite LaNi$_5$/C (a), LaNi$_5$/Cu (b), and LaNi$_5$/Pd (c) materials.

Figure 12.18 shows the discharge capacities as a function of the cycle number for some of the studied nanocomposite LaNi$_5$-type materials. Mechanical coating with graphite or palladium of LaNi$_5$-type phase and with palladium or cooper of LaNi$_5$-type phase effectively reduced the degradation rate of the studied electrode materials. Graphite inhibits the formation of new oxide layer on the surface of materials once the native oxide layer is broken during the ball-milling process.

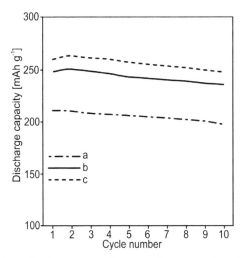

Figure 12.18 The discharge capacity as a function of cycle number for MA and annealed LaNi$_{3.75}$Mn$_{0.75}$Al$_{0.25}$Co$_{0.25}$/Cu (a), LaNi$_{3.75}$Mn$_{0.75}$Al$_{0.25}$Co$_{0.25}$/C (b), and LaNi$_{3.75}$Mn$_{0.75}$Al$_{0.25}$Co$_{0.25}$/Pd (c), (current density 40 mA g^{-1}, solution 6 M KOH, temperature 20°C).

References

1. Pan, Y. Y., and Nash, P. (1991). Lanthanum-Nickel, Phase diagrams of binary nickel alloys, Nash, P., ed., ASM International, Materials Park, OH, pp. 183–188.

2. Majchrzycki, W. (2002). The influence of microstructure on electrochemical properties of some alloys in Ni-MH applications. PhD thesis, Poznan University of Technology.

3. Zhang, F. L., Luo, Y. C., Chen, J. P., Yan, R. X., and Chen, J. H. (2007). La-Mg-Ni ternary hydrogen storage alloys with Ce$_2$Ni$_7$-type and Gd$_2$Co$_7$-type structure as negative electrodes for Ni/MH batteries. *J. Alloys Comp.* **430**: 302–307.

4. Chai, Y. J., Sakaki, K., Asano, K., Enoki, H., Akiba, E., and Kohno, T. (2007). Crystal structure and hydrogen storage properties of La-Mg-Ni-Co alloy with superstructure. *Scrip. Mater.* **57**: 545–548.

5. Liu, J., Han, S., Li, Y., Yang, S., Shen, W., Zhang, L., and Zhou, Y. (2013). An investigation on phase transformation and electrochemical properties of as-cast and annealed La$_{0.75}$Mg$_{0.25}$Ni$_x$ (x = 3.0, 3.3, 3.5, 3.8) alloys. *J. Alloy. Comp.* **552**: 119–126.

6. Yamamoto, T., Inui, H., Yamaguchi, M., Sato, K., Fujitani, S., Yonezu, I., and Nishio, K. (1997). Microstructures and hydrogen absorption/desorption properties of La–Ni alloys in the composition range of La-77.8–83.2 at% Ni. *Acta Mater.* **45**: 5213–5221.

7. Van Vucht, J. H. N., Kuijpers, F. A., and Burning, H. C. A. M. (1970). Reversible room-temperature absorption of large quantities of hydrogen by intermetallic compounds. *Philips Res. Rep.* **25**: 133–140.

8. Percheron-Guegan, A., Lartigue, G., Achard, J. C., Germi, P., and Tassett, F. (1980). Neutron and X-ray diffraction profile analyses and structure of LaNi$_5$, LaNi$_{5-x}$Al$_x$ and LaNi$_{5-x}$Mn$_x$ intermetallics and their hydrides (deuterides). *J. Less-Common Met.* **74**: 1–12.

9. Wallace, W. E., and Malik, S. K. (1978). Structure and bonding in metal hydrides. In *Hydrides for Energy Storage. Proc. Inter. Symp. Geilo*, Norway, 14–19 August 1977, pp. 33–42.

10. Sakai, T., Miyamura, H., Kuriyama, N., Ishikawa, H., and Uehara, L. (1994). Hydrogen storage for nickel-metal hydride battery. *Z. Phys. Chem.*, **183**: 333–346.

11. van Vucht, J. H. N., Kuujpers, F. A., and Bruning, H. C. A. M. (1970). Reversible room-temperature absorption of large quantities of hydrogen by intermetallic compounds. *Philips Res. Rep.*, **25**: 133–140.

12. Sakai, T., Natsuoka, M., and Iwakura, C. (1995). Rare earth intermetallics for metal hydrogen batteries. In Gschneider Jr., K. A., and Eyring, L. eds., *Handbook on the Physics and Chemistry of Rare Earth*, vol. 21, chapter 142, Amsterdam: Elsevier Science B.V., 1995. pp 135–180.

13. Fujii, H., Munehiro, S., Fujii, K., et al. (2002). Effect of mechanical grinding under Ar and H$_2$ atmospheres on structural and hydriding properties in LaNi$_5$. *J. Alloys Comp.* **330–332**: 747–751.

14. Schlapbach, L., and Züttel, A. (2001). Hydrogen-storage materials for mobile applications. *Nature* **414**: 353–358.

15. Bowman, R. C., Luo, C. H., Ahn, C. C., Witham, C. K., Fultz, B. (1995). The effect of tin on the degradation of LaNi$_{5-y}$Sn$_y$ metal hydrides during thermal cycling. *J. Alloys Comp.* **217**: 185–192.

16. Corre, S., Bououdina, M., Fruchart, D., and Adachi, G. (1998). Stabilisation of high dissociation pressure hydrides of formula $La_{1-x}Ce_xNi_5$ with carbon monoxide. *J. Alloys Comp.* **275–277**: 99–104.

17. Li, C., Wang, X., Li, X., and Wang, C. (1998). The hydrogen storage properties of $MnNi_{3.8}Co_{0.6}Mn_{0.55}Ti_{0.05}$ compounds prepared by conventional and melt-spinning techniques. *Electrochim. Acta* **43**: 1839–1818.

18. Geng, M., Han, J., Feng, F., and Northwood, D. O. (1999). Charging/discharging stability of a metal hydride battery electrode. *J. Electrochem. Soc.*, **146**: 2371–2375.

19. Tliha, M., Mathlouthi, H., Lamloumi, J., and Percheron-Guegan, A. (2007). AB_5-type hydrogen storage alloy used as anodic materials in Ni-MH batteries. *J. Alloys Comp.* **436**: 221–225.

20. Lin, J., Cheng, Y., Liang, F., Sun, L., Yin, D., Wu, Y., and Wang, L. (2014). High temperature performance of $La_{0.6}Ce_{0.4}Ni_{3.45}Co_{0.75}Mn_{0.7}Al_{0.1}$ hydrogen storage alloy for nickel/metal hydride batteries. *Int. J. Hydrog. Energy* **39**: 13231–13239.

21. Reilly, J. J., Adzic, G. D., Johnson, J. R., Vogt, T., Mukerjee, S., and MacBreen, J. (1999). The correlation between composition and electrochemical properties of metal hydride electrodes. *J. Alloys Comp.*, **293–295**: 569–582.

22. Yasuda, K. (1997). Effects of the materials processing on the hydrogen absorption properties of $MnNi_5$ type alloys. *J. Alloys Comp.*, **253–254**: 621–625.

23. Züttel, A., Chartouni, D., Gross, K., Spatz, P., Bächler, M., Lichtenberg, F., Fölzer, A., and Adkins, N. J. E. (1997). *Relationship between composition, volume expansion and cyclic stability of AB5-type metal hydride electrodes. J. Alloys Comp.*, **253–254**: 626–628.

24. Popovic, M. M., Grgur, B. N., Vojnovic, M. V., Rakin, P., and Krstajic, N. V. (2000). Electrochemical studies on $LaNi_{4.15}Co_{0.43}Mn_{0.40}Fe_{0.02}$ metal hydride alloy. *J. Alloys Comp.*, **298**: 107–113.

25. Lamloumi, J., Lartigue, C., Percheron Guegan, A., and Achard, J. J. (1980). In *Rare Earths in Modern Science and Technology*, MacCarthy, G. J., Rhyne, J. J., and Silver, H. B., eds., Plenum, New York.

26. Solonin, Y. M., Savin, V. V., Solonin, S. M., Skorokhod, V. V., Kolomiets, L. L., and Bratanich, T. I. (1997). Gas atomized powders of hydride-forming alloys and their application in rechargeable batteries. *J. Alloys Comp.*, **253–254**: 594–597.

27. Adzic, G. A., Johnson, J. R., Mukerjee, S., McBreen, J., and Reilly, J. J. (1997). Function of cobalt in AB_5H_x electrodes. *J. Alloys Comp.*, **253–254**: 579–582.

28. Lichtenberg, F., Köhler, U., Fölzer, A., Adkins, N. J. E., and Züttel, A. (1997). Development of AB_5 type hydrogen storage alloys with low Co content for rechargeable Ni–MH batteries with respect to electric vehicle applications. *J. Alloys Comp.* **253–254**: 569–573.

29. Hu, W. K., Lee, H., Kim, D. M., Jeon, S. W., and Lee, J. Y. (1998). Electrochemical behaviors of low-Co Mm-based alloys as MH electrodes. *J. Alloys Comp.* **268**: 261–265.

30. Ye, S. H., Gao, X. P., Liu, J., Wang, W. H., Yuan, H. T., Song, D. Y., and Zhang, Y. S. (1999). Characteristics of mixed hydrogen storage electrode. *J. Alloys Comp.* **292**: 191–198.

31. Hu, W. K. (2000). Improvement in capacity of cobalt-free Mm-based hydrogen storage alloys with good cycling stability. *J. Alloys Comp.*, **297**: 206–210.

32. Van Mal, H. H., Buschow, K. H. J., and Miedema, A. R. (1974). Hydrogen absorption in $LaNi_5$ and related compounds: Experimental observations and their explanation. *J. Less Common Metals* **35**: 65–76.

33. Takeshita, T., Dublon, G., McMasters, O. D., and Gschneidner, K. A. Jr. (1982). In *Rare Earths in Modern Science and Technology*, vol. 3, MacCarthy, G. J., Rhyne, J. J., and Silver, H. B., eds., Plenum, New York.

34. Yang, F. H., Cao, X. X., Zhang, Z. X., Bao, Z. W., Wu, Z., and Serge, N. N. (2012). Assessment on the long term performance of a $LaNi_5$ based hydrogen storage system. *Energy Procedia* **29**: 720–730.

35. Nahm, K. S., Kim, W. Y., Hong, S. P., and Lee, W. Y. (1992). The reaction kinetics of hydrogen storage in $LaNi_5$. *Int. J. Hydrogen Energy* **17**: 333–338.

36. Cheng, H. H., Yang, H. G., Lim S. L., Deng, X. X., Chen, D. M., and Yang, K. (2008). Effect of hydrogen absorption/desorption cycling on hydrogen storage performance of $LaNi_{4.25}Al_{0.75}$. *J. Alloys Comp.* **453**: 448–52.

37. Young K.-H., and Nei J., (2013). The current status of hydrogen storage alloy development for electrochemical applications, *Materials* **6**: 4574–4608.

38. Iwakura, C., Fukuda, K., Senoh, H., Inoue, H., Matsuoka, M., and Yamamoto, Y. (1998). Electrochemical characterization of $MmNi_{4.0-x}Mn_{0.75}Al_{0.25}Co_x$ electrodes as a function of cobalt content. *Electrochim. Acta* **43**: 2041–2046.

39. Balogun, M., Wang, Z., Chen, H., Deng, J., Yao, Q., and Zhou, H. (2013). Effect of Al content on structure and electrochemical properties of LaNi$_{4.4-x}$Co$_{0.3}$Mn$_{0.3}$Al$_x$ hydrogen storage alloys. *Int. J. Hydrog. Energy* **39**: 10926–10931.

40. Giza, K., Bala, H., Drulis, H., Hackemer, A., and Folcik, L. (2012). Gaseous and electrochemical hydrogenation properties of LaNi$_{4.3}$(Co, Al)$_{0.7-x}$In$_x$ alloys. *Int. J. Electrochem. Sci.* **7**: 9881–9891.

41. Wang, W., Chen, Y., Li, Q., and Yang, W. (2013). Microstructures and electrochemical properties of LaNi$_{3.8-x}$Al$_x$ hydrogen storage alloys. *J. Rare Earths* **32**: 497–501.

42. Nowak, M. (2003). Influence of Al, Mn, Co substitution for Ni for the structure and electrochemical properties of nanocrystalline LaNi$_5$-type alloys. PhD thesis, Poznan University of Technology.

43. Chao, D., Zhong, C., Ma, Z., Yang, F., Wu, Y., Zhu, D., Wu, C., and Chen, Y. (2012). Improvement in high-temperature performance of Co-free high-Fe AB$_5$-type hydrogen storage alloys. *Int. J. Hydrog. Energy* **37**: 12375–12383.

44. Yang, S., Han, S., Li, Y., Yang, S., and Hu, L. (2011). Effect of substituting B for Ni on electrochemical kinetic properties of AB$_5$-type hydrogen storage alloys for high-power nickel/metal hydride batteries. *Mater. Sci. Eng. B* **176**: 231–236.

45. Srivastava, S., and Upadhyay, R. K. (2010). Investigations of AB$_5$-type negative electrode for nickel-metal hydride cell with regard to electrochemical and microstructural characteristics. *J. Power Sources* **195**: 2996–3001.

46. Jurczyk, M., Nowak, M., Jankowska, E., and Jakubowicz, J. (2002). Structure and electrochemical properties of the mechanically alloyed La(Ni,M)$_5$ materials. *J. Alloys Comp.* **339**: 339–343.

47. Smardz, L., Smardz, K., Jurczyk, M., and Nowak, M. (2001) Structure and electronic properties of La(Ni,Al)$_5$ alloys. *Cryst. Res. Technol.* **36**: 1385–1392.

48. Szajek, A., Smardz, L., Jurczyk, M., Smardz, K., and Nowak, M. (2002) Electronic properties of LaNi$_5$-type alloys. *Czech. J. Phys.* **52**: A209–A212.

49. Jurczyk, M., Smardz, K., Rajewski, W., and Smardz, L. (2001). Nanocrystalline LaNi$_{4.2}$Al$_{0.8}$ prepared by mechanical alloying and annealing and its hydride formation. *Mater. Sci. Eng. A* **303**: 70–76.

50. Smardz, L., Jurczyk, M., Smardz, K., and Nowak, M. (2002). Segregation effect on nanocrystalline La(Ni,Co,Al)$_5$ surface. *Czech. J. Phys.* **52**: A177–A180.

51. Jurczyk, M., Jankowska, E., Nowak, M., and Jakubowicz, J. (2002). Nanocrystalline titanium type metal hydrides prepared by mechanical alloying. *J. Alloys Comp.* **336**: 265–269.
52. Jurczyk, M., Jankowska, E., Nowak, M., and Wieczorek, I. (2003). Electrode characteristics of nanocrystalline TiFe-type alloys. *J. Alloys Comp.* **354**: L1–L4.
53. Jurczyk, M., Smardz, L., Smardz, K., Nowak, M., and Jankowska, E. (2003). Nanocrystalline LaNi$_5$-type electrode materials for Ni-MH$_x$ batteries. *J. Solid State Chem.* **171**: 30–37.
54. Jurczyk, M., Nowak, M., and Jankowska, E. (2002). Nanocrystalline LaNi$_{4-x}$Mn$_{0.75}$Al$_{0.25}$Co$_x$ electrode materials produced by mechanical alloying (0 ≤ x ≤ 1.0). *J. Alloys Comp.* **340**: 281–285.
55. Szajek, A., Jurczyk, M., and Rajewski, W. (2000). The electronic and electrochemical properties of the LaNi$_5$, LaNi$_4$Al and LaNi$_3$AlCo systems. *J. Alloys Comp.* **307**: 290–296.
56. Andersen, O. K. (1975). Linear methods in band theory. *Phys. Rev. B* **12**, 3060–3083.
57. Szajek, A., Jurczyk, M., Smardz, L., Okonska, I., and Jankowska, E. (2007). Electrochemical and electronic properties of nanocrystalline Mg-based hydrogen storage materials. *J. Alloys Comp.* **436**: 345–350.
58. Ming, B. I., and Jang, Y. R. (1991). The effect of the spin–orbit interaction on the electronic structure of magnetic materials. *J. Phys. Condens. Matter.* **3**: 5131–5561.
59. Perdew, J. P., Chevary, J. A., Vosko, S. H., Jackson, K. A., Pederson, M. R., Singh, D. J., Fiolhais, C. (1992). Atoms, molecules, solids, and surfaces: Applications of the generalized gradient approximation for exchange and correlation. *Phys. Rev. B* **46**: 6671–6687.
60. Andersen, O. K., Jepsen, O., Šob, M. (1987). In: Yussouff, M. S., ed., *Electronic Structure and Its Applications*, Springer, Berlin.
61. Blöchl, P., Jepsen, O., and Andersen, O. K. (1994). Improved tetrahedron method for Brillouin-zone integrations. *Phys. Rev. B* **49**: 16223–16233.
62. Gurewitz, E., Pinto, H., Dariel, M., and Shaked, H. (1983). Neutron diffraction study of LaNi$_4$Co and LaNi$_4$CoD$_4$. *J. Phys. F Met. Phys.* **13**: 545–554.
63. Joubert, J. M., Černy, R., Latroche, M., Percheron-Guégan, A., and Yvon, K. (1998). Site occupancies in the battery electrode material LaNi$_{3.55}$Mn$_{0.4}$Al$_{0.3}$Co$_{0.75}$ as determined by multiwavelength synchrotron powder diffraction. *J. Appl. Crystallogr.* **31**: 327–332.

64. Sluiter, M., Takahashi, M., and Kawazoe, Y. (1997). Theoretical study of phase stability in LaNi$_5$-LaCo$_5$ Alloys. *J. Alloys Comp.* **248**: 90–97.
65. van Mal, H. H., Buschow, K. H. A., and Kuijpers, F. A. (1973). Hydrogen absorption and magnetic properties of LaCo$_{5x}$Ni$_{5-5x}$ compounds. *J. Less Common Met.* **32**: 289–296.
66. Jurczyk, M., Nowak, M., Okonska, I., Smardz, L., and Szajek, A. (2009). Nanocomposite hydride LaNi$_5$/A- and Mg$_2$Ni/A-type materials (A = C, Cu, Pd). *Mat. Sci. Forum* **610–613**: 472–479.

Chapter 13

Mg-3d-Based Hydrogen Storage Alloys

Marek Nowak and Mieczyslaw Jurczyk

Institute of Materials Science and Engineering,
Poznan University of Technology, Poznan, Poland

marek.nowak@put.poznan.pl, mieczysław.jurczyk@put.poznan.pl

13.1 Introduction

Magnesium-based alloys have been extensively studied during last years, but the microcrystalline Mg_2Ni alloy can reversibly absorb and desorb hydrogen only at high temperatures [1–5]. Substantial improvements in the hydriding-dehydriding properties of Mg-type metal hydrides could possibly be achieved by formation of nanocrystalline structures [4, 6, 7]. It was found that the electrochemical activity of nanocrystalline hydrogen storage alloys can be improved in many ways, by alloying with other elements [8, 9], by ball-milling the alloy powders with a small amount of nickel or graphite powders [10–13]. For example, the surface modification of nanocrystalline hydrogen storage alloys with graphite by ball milling leads to an improvement in both discharge capacity and charge-discharge cycle life [13, 14].

Recently, nanocrystalline and microcrystalline Mg_2Cu, $(Mg_{1-x}M_x)_2Ni$ alloys, as well as Mg_2Cu/M' and $(Mg_{1-x}M_x)_2Ni/M'$,

---

Handbook of Nanomaterials for Hydrogen Storage
Edited by Mieczyslaw Jurczyk
Copyright © 2018 Pan Stanford Publishing Pte. Ltd.
ISBN 978-981-4745-66-6 (Hardcover), 978-1-315-36444-5 (eBook)
www.panstanford.com

(x = 0, 0.5; M = Al, Mn; M′ = C, Pd) nanocomposites have been prepared by mechanical alloying followed by annealing, by a diffusion method and by ball milling, respectively [15–21]. Formation of nanocrystalline alloys was achieved by annealing the amorphous materials in high-purity argon atmosphere.

13.2 Mg–Ni and Mg–Cu Phase Diagrams

As shown in Fig. 13.1 magnesium forms two intermetallic compounds with nickel, namely Mg$_2$Ni (space group $P6_222$) and MgNi$_2$ (space group $P6_3/mmc$) [22]. Mg$_2$Ni reacts with hydrogen to form a ternary hydride, Mg$_2$NiH$_4$. The MgNi$_2$ phase does not react with hydrogen at pressures up to 540 atm in the temperature range –196 to 300°C. The hydrogen content in Mg$_2$NiH$_4$ is about 3.6 wt%. Many substituted Mg$_2$Ni alloys have been studied to improve their hydrogenation properties.

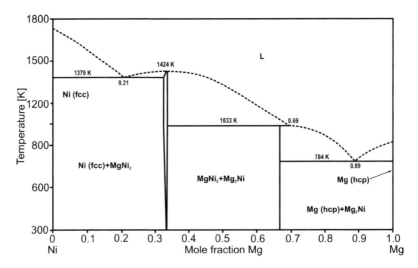

Figure 13.1 Mg–Ni phase diagram.

As shown in Fig. 13.2, magnesium forms two intermetallic compounds with copper, namely Mg$_2$Cu (space group $Fddd$) and MgCu$_2$ (space group $Fd3m$) [23]. Limited solubility of Cu in Mg as well as Mg in Cu has been reported. The solubility of Cu in Mg increases from about 0.1 at% Cu at room temperature to about 0.4–0.5 at% Cu at 485°C (Fig. 13.2).

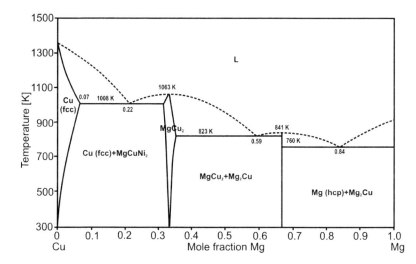

Figure 13.2 Mg–Cu phase diagram.

13.3 Mg₂Ni-Type Alloys

Figure 13.3 shows a series of XRD spectra of mechanically alloyed 2Mg–Ni powder mixture (0.453 wt% Mg + 0.547 wt% Ni) subjected to milling in increasing time [25]. The nanostructured Mg_2Ni with broad diffraction peaks are already found after 5 h of MA process. The powder mixture milled for more than 30 h has transformed directly to a hexagonal-type phase (Fig. 13.3b). Finally, the obtained powder was heat treated in high-purity argon atmosphere at 450°C for 0.5 h to obtain the desired microstructure (Fig. 13.3c). All diffraction peaks were assigned to those of the hexagonal crystal structure with cell parameters a = 5.216 Å, c = 13.246 Å. Table 13.1 reports the cell parameters of all the studied materials. For x = 0.5 in $Mg_{2-x}M_xNi$ (M = Al or Mn) the crystalline phase of a CsCl-type cubic structure is formed [4, 24]. The average size of amorphous 2Mg–Ni powders, according to AFM studies, was of the order of 30 nm.

At room temperature, the original nanocrystalline alloy, Mg_2Ni, absorbs hydrogen, but almost does not desorb it. At temperatures above 250°C the kinetic of the absorption-desorption process improves considerably and for nanocrystalline Mg_2Ni alloy the reaction with hydrogen is reversible. The hydrogen content in this material at 300°C is 3.25 wt%. Upon hydrogenation, Mg_2Ni

transforms into the hydride Mg$_2$Ni-H phase. It is important to note that between 245–210°C the hydride Mg$_2$Ni-H phase transforms from a high temperature cubic structure to a low temperature monoclinic phase [1]. When hydrogen is absorbed by Mg$_2$Ni beyond 0.3 H per formula unit, the system undergoes a structural rearrangement to the stoichiometric complex Mg$_2$Ni-H hydride, with an accompanying 32% increase in volume.

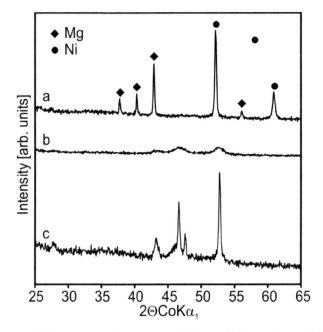

Figure 13.3 XRD spectra of a mixture of 2Mg and Ni powders: (a) initial state (elemental powder mixture), (b) after MA for 30 h, and (c) and heat treated at 450°C for 0.5 h.

The hydriding kinetics of the alloy is improved after substitution of some amounts of magnesium by Al or Mn (Table 13.1). The results show that the maximum absorption capacity reaches 3.25 wt% for pure nanocrystalline Mg$_2$Ni alloy. This is lower than the microcrystalline Mg$_2$Ni alloy (3.6 wt% [25, 26]) due to a significant amount of strain, chemical disorder and defects introduced into the material during the mechanical alloying process [6, 27]. The concentration of hydrogen in produced nanocrystalline Mg$_{2-x}$M$_x$Ni (M = Al, Mn) alloys strongly decreases with increasing M contents [24].

Table 13.1 Structure, lattice parameters, discharge capacities, and hydrogen contents for nanocrystalline Mg$_2$Ni-type materials

Alloy	Structure and lattice constants	Discharge capacity [mA h g^{-1}]	Hydrogen content at 300°C [wt%]
Mg$_2$Ni nanocrystalline	hexagonal a = 5.216 Å, c = 13.246 Å	100	3.25
Mg$_{1.75}$Mn$_{0.25}$Ni nanocrystalline	Hexagonal a = 5.185 Å, c = 13.097 Å	148	2.50
Mg$_{1.5}$Mn$_{0.5}$Ni nanocrystalline	cubic a = 3.137 Å	241	0.65
Mg$_{1.75}$Al$_{0.25}$Ni nanocrystalline	hexagonal a = 5.193 Å, c = 13.173 Å	105	1.75
Mg$_{1.5}$Al$_{0.5}$Ni nanocrystalline	cubic a = 3.149 Å	175	0.26
Mg$_2$Ni microcrystalline	hexagonal a = 5.223 Å, c = 13.30 Å	—	3.6

Note: Data for parent microcrystalline Mg$_2$Ni alloy were also included for comparison.

For example, the concentration of hydrogen in produced nanocrystalline Mg$_2$Ni alloys strongly decreases with increasing Mn contents. The hydrogen content at 300°C in nanocrystalline Mg$_{1.5}$Mn$_{0.5}$NiH was only 0.65 wt% (Table 13.1, Fig. 13.4).

The Mg$_2$Ni electrode, mechanically alloyed and annealed, displayed the maximum discharge capacity (100 mA h g^{-1}) at the first cycle but degraded strongly with cycling (Fig. 13.5). The poor cyclic behavior of Mg$_2$Ni electrodes is attributed to the formation of Mg(OH)$_2$ on the electrodes, which has been considered to arise from the charge-discharge cycles [28]. To avoid the surface oxidation, we have examined the effect of magnesium substitution by Al and Mn in Mg$_2$Ni-type material. This alloying greatly improved the discharge capacities (Table 13.1). In nanocrystalline

Mg$_{1.5}$Mn$_{0.5}$Ni alloy discharge capacities up to 241 mA h g^{-1} was measured. A similar phenomenon to that described here has been observed by Yuan et al. in Mg$_{2-x}$Al$_x$Ni-type powders [29].

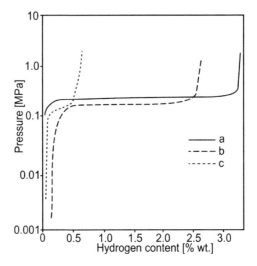

Figure 13.4 PC isotherms at 300°C of hydrogen desorption from nanocrystalline Mg$_{2-x}$Mn$_x$Ni-H alloys; (a) $x = 0$, (b) $x = 0.25$ and (c) $x = 0.5$.

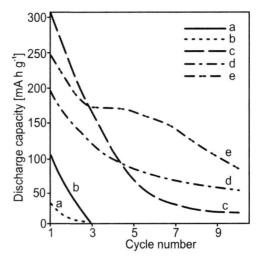

Figure 13.5 The discharge capacity as a function of cycle number for MA and annealed Mg$_2$Cu (a), Mg$_2$Ni (b) and Mg$_2$Ni/Pd (c), Mg$_{1.5}$Mn$_{0.5}$Ni/C (d), and Mg$_{1.5}$Al$_{0.5}$Ni/Pd (e) electrodes (solution, 6M KOH; temperature, 20°C).

It is important to note that the surface chemical composition of nanocrystalline Mg_2Ni-type alloy studied by X-ray photoelectron spectroscopy (XPS) showed the strong surface segregation under UHV conditions of Mg atoms in the MA nanocrystalline Mg_2Ni alloy [30]. This phenomenon could considerably influence the hydrogenation process in such a type of materials, too. The same surface segregation was observed earlier by Stefanov in Mg–Ni films, where magnesium was found to segregate at the surface [31].

Independently, surface chemical composition of $Mg_{1.8}Ag_{0.2}Ni$ alloy was studied by XPS method [32]. The result shows that the surface of the alloy was covered by an amorphous layer, which contained much oxygen. The amorphous layer on the surface of this alloy was MgO which formed due to the evaporation of Mg and oxidation. Additionally, it was demonstrated that the component distribution in an amorphous Mg–Ni alloy, prepared by MA, is depth-dependent [33]. At the top surface, Mg existed as oxide and dominates, while underneath the top surface, a concentrated metallic Ni layer was present. The lower lying Ni atoms formed metallic subsurface layer and are responsible for the observed high hydrogenation rate in accordance with earlier findings [34, 35]. The oxidation process is depth limited such that an oxide-covering layer with a well-defined thickness is formed by which the lower lying metal is prevented from further oxidation. In this way one can obtain a self-stabilized oxide–metal structure. This Ni-enriched layer not only serves an effective catalyst but also facilitates the kinetics of hydrogen absorption.

The experimental XPS valence bands measured for the nanocrystalline MA Mg_2Ni alloy and microcrystalline Mg_2Ni thin film are shown in Fig. 13.6. The experimental XPS valence bands measured for nanocrystalline Mg_2Ni alloy (Fig. 13.6a) showed a significant broadening compared to that obtained for the microcrystalline thin film (Fig. 13.6c). The substitution of Mg by Mn in the nanocrystalline Mg_2Ni alloy causes a further band broadening (Fig. 13.6b). On the other hand, our XPS measurements showed that the substitution of Mg by Al or Mn leads only to a small valence band broadening (not shown in Fig. 13.6). The reasons responsible for the band broadening of the nanocrystalline Mg_2Ni-type alloys are probably associated with a strong deformation of the nanocrystals in the MA samples [35]. Normally the interior of the nanocrystal is constrained and the distances between atoms located at the grain boundaries expanded [36]. Furthermore, the Al and Mn atoms could also occupied metastable positions in the nanocrystals. The strong modifications of the electronic structure of the nanocrystalline Mg$_2$Ni-type alloy could significantly influence on its hydrogenation properties, similarly to the behavior observed earlier for the nanocrystalline FeTi- and LaNi$_5$-type alloys and Mg$_2$Cu [21, 35].

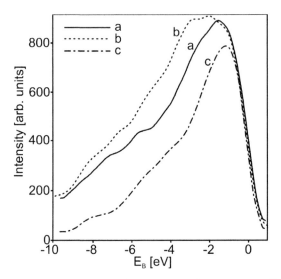

Figure 13.6 XPS valence band (Al-Kα) spectra for nanocrystalline Mg$_2$Ni (a) and Mg$_{1.5}$Mn$_{0.5}$Ni (b) alloys and microcrystalline Mg$_2$Ni thin films (c). The XPS measurements were performed immediately after heating in UHV conditions followed by removing of a native oxide and possible impurities layer using ion gun etching system (see text).

13.4 Mg$_2$Cu-Type Alloys

Figure 13.7 shows a series of XRD spectra of mechanically alloyed 2Mg–Cu powder mixture (0.433 wt% Mg + 0.567 wt% Cu) subjected to milling in increasing time [37]. The powder mixture milled for more than 18 h has transformed directly to an orthorhombic-type phase. Finally, the obtained powder was heat treated in high-purity argon atmosphere at 450°C for 0.5 h to obtain the desired microstructure. All diffraction peaks were assigned to those of orthorhombic-type structure with cell parameters a = 9.119(4) Å, b = 18.343(4) Å, c = 5.271(1) Å (Fig. 13.7d).

Table 13.2 reports the cell parameters of all the studied materials. According to the Scherrer method for XRD profiles, the average size of 2Mg–Cu mechanically alloyed for 40 h powders was of the order of 30 nm; after heat treatment in high-purity argon atmosphere at 450°C for 0.5 h the mean crystallite size of the nanocrystalline alloy estimated from AFM experiment was about 50 nm.

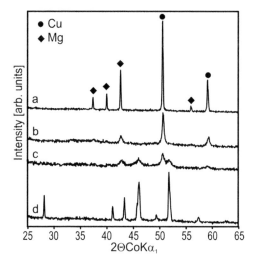

Figure 13.7 XRD spectra of a mixture of 2Mg and Cu powders mechanically alloyed for different times in an argon atmosphere: initial state (elemental powder mixture) (a), after MA for 4.5 h (b), after MA for 18 h (c) and heat treated at 450°C for 0.5 h (d).

At room temperature, the original nanocrystalline Mg_2Cu alloy absorbs hydrogen, but almost does not desorb it. At temperatures above 250°C the kinetic of the absorption-desorption process improves considerably and for nanocrystalline Mg_2Cu alloy the reaction with hydrogen is reversible. Upon hydrogenation, Mg_2Cu transforms into the hydride MgH_2 + $MgCu_2$ phases. At 300°C, the maximum absorption capacity reaches 2.25 wt% for pure nanocrystalline Mg_2Cu alloy. This is lower than in the microcrystalline Mg_2Cu alloy (2.6 wt%) because of a significant amount of strain, chemical disorder, and defects introduced into the material during the mechanical alloying process [7].

The Mg_2Cu electrode, mechanically alloyed and annealed, displayed the maximum discharge capacity (26.5 mA h g^{-1}) at the first cycle but degraded strongly with cycling (see Table 13.2, Fig. 13.4). The poor cyclic behavior of Mg$_2$Cu electrodes is attributed to the formation of Mg(OH)$_2$ on the electrodes, which has been considered to arise from the charge-discharge cycles [28]. The discharge capacity of coated nanocrystalline Mg$_2$Cu powders with palladium was improved. Mechanical coating with palladium effectively reduced the degradation rate of the studied electrode material.

Table 13.2 Structure, lattice parameters and discharge capacities for nanocrystalline and nanocomposite Mg$_2$Cu-type materials

Material	Structure and lattice constants [Å]	Discharge capacity [mA h/g] 1st cycle	3rd cycle
Nanocrystalline Mg$_2$Cu	Orthorhombic a = 9.119, b = 18.343, c = 5.271	26.5	4.7
Nanocomposite Mg$_2$Cu/C	Orthorhombic/hexagonal a = 9.046, b = 18.463, c = 5.274/a = 2.456, c = 6.994	26.0	18.5
Nanocomposite Mg$_2$Cu/Pd	Orthorhombic/cubic a = 9.046, b = 18.463, c = 5.274/a = 3.890	26.3	19.3

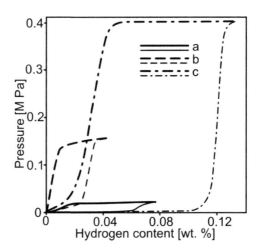

Figure 13.8 Electrochemical pressure-composition isotherms for absorption (solid line) and desorption (dashed line) of hydrogen on microcrystalline Mg$_2$Cu (a), nanocrystalline Mg$_2$Cu (b) and nanocomposite Mg$_2$Cu/Pd (c); (6M KOH solution; the charge/discharge conditions were 4 mA g^{-1} and the cut-off potential was –0.700 V).

In the case of Mg$_2$Cu-type alloys, the capacity of the alloy electrode in relation to the amount of absorbed hydrogen (wt%) was calculated based on the input/output charge. The e.p.c. isotherms determined on the studied Mg$_2$Cu-type materials are illustrated in Fig. 13.8. The isotherms show an increase of the equilibrium hydrogen pressure and an increase in the amount of hydrogen observed for the nanocomposite Mg$_2$Cu/Pd material (curve c) in comparison with the microcrystalline (curve a) and nanocrystalline (curve b) alloys.

Application of Mg-based alloys focused attention also on the electronic structure of Mg$_2$Cu-type hydrogen storage materials [20]. The experimental XPS valence band spectrum measured for the microcrystalline Mg$_2$Cu alloy is shown in Fig. 13.9. The main XPS signal is originated from the Cu(3d) electrons. Furthermore, it has been found a good agreement between the shape and position of the experimentally determined spectrum (Fig. 13.9) and theoretically calculated XPS valence band [38]. Note that the theoretical band calculation was performed for ideal single-crystalline Mg$_2$Cu alloy. Therefore, some disorder as well as Cu and Mg atoms located near grain boundary of the microcrystalline sample could be responsible for a small broadening of the experimental XPS spectrum for the microcrystalline Mg$_2$Cu alloy. Figure. 13.9 shows the experimental XPS valence bands measured for MA nanocrystalline Mg$_2$Cu alloy. The results showed a significant broadening of the valence band measured for the MA nanocrystalline sample compared to that obtained for the microcrystalline Mg$_2$Cu. The reasons responsible for the band broadening of the nanocrystalline Mg$_2$Cu alloy are probably associated with a strong deformation of the nanocrystals in the MA samples [35].

The main modification of the electronic structure takes place for binding energies between the bottom of valence band and the main peak of 3d element, about 5–9 eV below the Fermi level. Furthermore, band structure calculations for Mg$_2$NiH$_4$ [39, 40] showed that electronic states of H are located in this region of the spectrum. Additional states in nanocrystalline structure, manifested by increasing of the density of states, can lead to stronger hybridization between the hydrogen states and states of the nanocrystalline matrix. Thereby hydrogen in nanocrystalline structure can be stronger bonded than in the microcrystalline one.

Therefore, the strong modifications of the electronic structure of the nanocrystalline Mg$_2$Cu (Fig. 13.9) alloys could significantly influence on its hydrogenation properties.

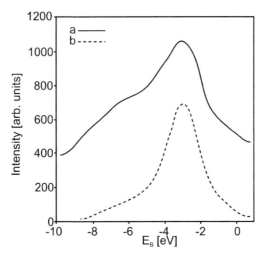

Figure 13.9 XPS valence band (Al-Kα) spectra for nanocrystalline Mg$_2$Cu (a) and microcrystalline Mg$_2$Cu (b) alloys. The XPS measurements were performed immediately after heating in UHV conditions followed by removing of a native oxide and possible impurities layer using ion gun etching system.

13.5 Effect of Ball-Milling with Graphite and Palladium

In order to improve the electrochemical properties of the studied nanocrystalline electrode materials, the ball-milling technique was applied to the Mg-based alloys using the graphite and palladium as a surface modifiers (Fig. 13.10). These composite materials with 10 wt% of C or Pd were produced by ball-milling for 1 h. Ball-milling with graphite is responsible for a sizeable reduction of the crystallite sizes of Mg$_{1.5}$Mn$_{0.5}$Ni/C from 30 to 20 nm.

Figure 13.5 shows the discharge capacities as a function of the cycle number for studied nanocomposite materials. When coated with graphite or palladium, the discharge capacity of nanocrystalline Mg$_{1.5}$Mn$_{0.5}$Ni and Mg$_{1.5}$Al$_{0.5}$Ni powders was increased. The elemental graphite was distributed on the surface of ball milled alloy particles homogenously and role of these particles is to catalyze the dissociation of molecular hydrogen on the surface of studied alloy. Mechanical coating with graphite and palladium effectively reduced the degradation rate of the studied electrode materials. Compared to that of the uncoated powders, the degradation of the coated was suppressed. Recently, Iwakura et al. [41] have demonstrated that the modification of graphite on the MgNi alloy in the MgNi–graphite composite is mainly a surface one. Raman and XPS investigations indicated the interaction of graphite with MgNi alloy occurred at the Mg part in the alloy. Graphite inhibits the formation of new oxide layer on the surface of materials once the native oxide layer is broken during the ball-milling process.

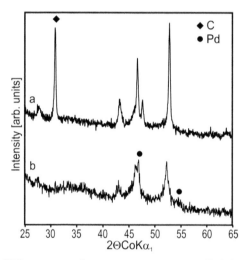

Figure 13.10 XRD spectra of nanocomposite Mg$_2$Ni/C (a) and Mg$_2$Ni/Pd (b) materials.

The experimental XPS valence bands measured for the Mg$_2$Ni/Pd and Mg$_2$Cu/Pd composites are shown in Fig. 13.11. The experimental XPS valence bands measured for nanocomposite alloys showed a significant broadening compared to that obtained for the microcrystalline Mg$_2$Cu or Mg$_2$Ni alloys. The maximum of the nanocrystalline Mg$_2$Ni valence band spectrum is located about 1.78 eV closer to the Fermi level compared to that measured for the nanocomposite Mg$_2$Ni/Pd. Results also showed a significant broadening of the valence bands of studied nanocomposites compared to those obtained by theoretical band calculations.

Especially, a clear broadening of the band can be visible when compared experimental XPS valence band for the nanocrystalline Mg$_2$Cu alloy and nanocomposite Mg$_2$Cu/Pd material. The reasons responsible for the band broadening of the nanocrystalline Mg$_2$Ni and Mg$_2$Cu alloys are probably associated with a strong deformation of the nanocrystals in the MA samples [42]. Normally the interior of the nanocrystal is constrained and the distances between atoms located at the grain boundaries expanded. The valence band spectra of the MA samples could be also broadened due to an additional disorder introduced during formation of the nanocrystalline structure.

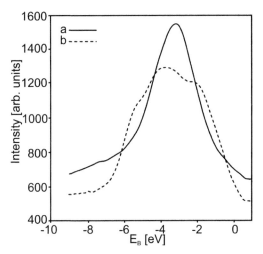

Figure 13.11 XPS valence band (Al-Kα) spectra for Mg$_2$Cu/Pd (a) and Mg$_2$Ni/Pd (b) nanocomposites. The XPS measurements were performed immediately after heating in UHV conditions followed by removing of a native oxide and possible impurities layer using ion gun etching system.

13.6 Amorphous 2Mg + 3d Alloys Doped by Nickel Atoms (3d = Fe, Co, Ni, Cu)

Recently, an amorphous 2Mg + 3d/x wt% Ni materials were prepared by mechanical alloying (MA) of Mg and 3d elemental powders (3d = Fe, Co, Ni, Cu; $0 \leq x \leq 200$) under high-purity argon atmosphere (Figs. 13.12 and 13.13) [17, 37, 43]. For example, in Table 13.3 the effect of the Ni addition on the electrochemical properties of the synthesized nanostructured 2Mg+Fe/x wt% Ni alloys was presented.

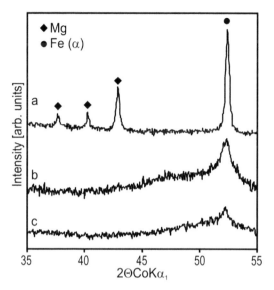

Figure 13.12 XRD spectra of a nanoscale 2Mg+Fe/x wt% Ni materials: x = 0 (a), x = 100 (b) and x = 200 (c).

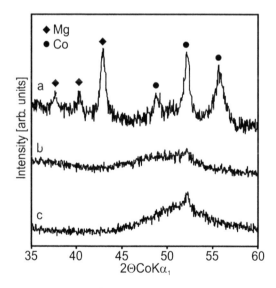

Figure 13.13 XRD spectra of a nanoscale 2Mg+Co/x wt% Ni materials: x = 0 (a), x = 100 (b) and x = 200 (c).

Table 13.3 Particle size, discharge capacities and hydrogen contents for amorphous 2Mg + Fe/x wt% Ni materials

Nickel content [wt%]—x	Particle size [nm]	Discharge capacity [mAh/g] 1st cycle	10th cycle	Hydrogen content at 300°C [wt%]
0	30 (Mg) 42 (α-Fe)	0	0	3.25
100	29	155	45	1.55
200	18	134	75	0.35

The behavior of MA process was studied by X-ray diffraction. Figures 13.12 and 13.13 show a series of XRD spectra of mechanically alloyed 2Mg + 3d/x wt% Ni powder mixtures for 48 h. In the case of nickel-free 2Mg + Fe and 2Mg + Co powder mixtures only Mg and α-Fe lines as well as Mg and Co lines are observed, respectively (Figs. 13.12a and 13.13a). On the other hand, 2Mg+Ni and 2Mg+Cu alloys except Mg and Ni or Mg and Cu contain a small amount of Mg_2Ni (d = 2.246 Å) and Mg_2Cu (d = 2.268 Å) phases, respectively. The nickel addition to 2Mg+3d powder mixtures is advantageous for the formation of amorphous structures (Figs. 13.12b,c and 13.13b,c). But differentiation between a "truly" amorphous, extremely fine grained or a material in which very small crystals are embedded in an amorphous matrix has not been easy on the basis of diffraction analysis.

The microstructure that forms during MA consists of layers of the starting material. The lamellar structure is increasingly refined during further mechanical alloying. After 48 h of mechanical alloying the sample shows cleavage fracture morphology and inhomogeneous size distribution. According to the Scherrer method for XRD profiles, the average size of 2Mg + Fe mechanically alloyed for 48 h powders was of the order of 30 and 42 nm for Mg and α-Fe, respectively (Table 13.3). On the other hand, for 2Mg+Fe/x wt% Ni powder mixtures MA for 48 h the average particle sizes were 29 and 18 nm for x = 100 and 200 wt% of Ni, respectively. Xiao at al. pointed that the ball-milled (2Mg+Fe/200 wt% Ni) may be amorphous 2Mg–Fe alloy which are inlaid with microcrystalline/nanocrystalline Ni particles homogeneously [44].

At room temperature, the mechanically alloyed 2Mg+3d alloys absorb hydrogen, but does not desorb it completely.

At temperatures above 250°C the kinetics of the absorption–desorption process improves considerably and for 2Mg+3d mixtures the reaction with hydrogen is reversible. For example, upon hydrogenation, 2Mg+3d mixtures transforms into the MgH_2 + 3d phases (see Fig. 13.14). On the other hand, upon hydrogenation, 2Mg+3d/100 wt% Ni and 2Mg+3d/200 wt% Ni mixtures transform into the $Mg_2(3d)H_6$ + 3d and MgH_2 + 3d, respectively. The concentration of hydrogen in amorphous 2Mg+3d/x wt% Ni materials strongly decreases with increasing nickel contents. In the case of 2Mg+Fe/x wt%, the hydrogen content at 300°C in these materials was 3.25, 1.55, and 0.35 wt% for x = 0, 100 and 200, respectively (see Fig. 13.15) [37, 43].

The cycling dischargeability of 2Mg + 3d/x wt% Ni electrode alloys (3d = Fe and Co) are shown in Figs. 13.16 and 13.17. The 2Mg + 3d electrodes, mechanically alloyed, displayed the zero discharge capacities, at 10th cycle. The discharge capacity of amorphous 2Mg+3d ball milled with Ni was improved. With the increasing of nickel content, the discharge capacity of the 2Mg+3d alloys increases first and then decreases reaching a maximum of 155, 93, 175 and 130 mAh g^{-1} at x = 100 for 3d = Fe, Co, Ni and Cu, respectively. Additionally, for 2Mg + Ni/Ni with increasing nickel, the discharge capacity increases from 90 mAh g^{-1} (x = 0) to 175 mAh g^{-1} (x = 100) and then decreases to 64 mAh g^{-1} (x = 200). For all 2Mg + 3d studied materials, mechanical alloying with nickel effectively reduced the degradation rate of the studied electrodes. The elemental nickel was distributed inside of mechanically alloyed Mg-3d particles homogenously. It is generally agreed that the role of nickel particles is to catalyze the electrochemical reaction and/or reduce the diffusion resistance of hydrogen. High energy ball milling of 2Mg+3d with nickel effectively reduced the degradation rate of the studied electrode material. The higher the Ni content is, the more resistant the alloy is to the corrosion of the alkaline electrolyte. Generally, the poor cyclic behavior of 2Mg+3d/Ni electrodes is attributed to the formation of $Mg(OH)_2$ on the electrodes, which has been considered to arise from the charge-discharge cycles [28]. Generally, the amorphous structure is easily formed with increasing Ni content in the alloys, and the amorphous alloys have good corrosion resistance property. This fact can improve the cycling stability of the alloys [44].

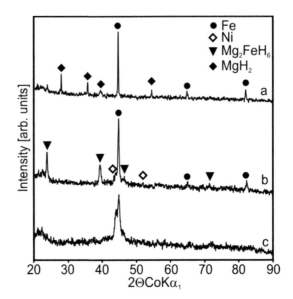

Figure 13.14 Room temperature XRD spectra of a mechanically alloyed 2Mg + Fe/x wt% Ni powders after hydrogenation at 300°C: x = 0 (a), x = 100 (b) and x = 200 (c).

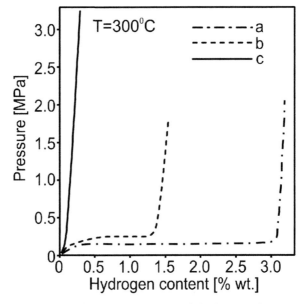

Figure 13.15 PC isotherms at 300°C of hydrogen desorption from 2Mg + Fe/x wt% Ni alloys: (a) x = 0, (b) x = 100 and (c) x = 200.

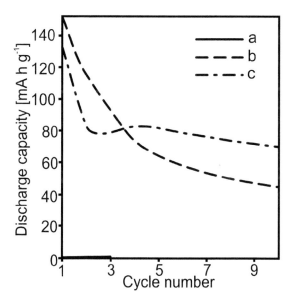

Figure 13.16 Discharge capacities of amorphous 2Mg + Fe/x wt% Ni: x = 0 (a), x = 100 (b) and x = 200 (c); (current density of charge/discharge was 4 mA g^{-1}).

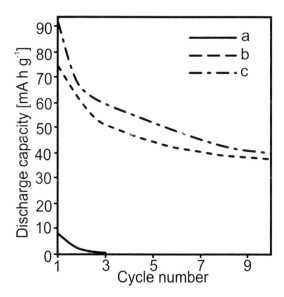

Figure 13.17 Discharge capacities of amorphous 2Mg + Co/x wt% Ni: x = 0 (a), x = 100 (b) and x = 200 (c); (current density of charge/discharge was 4 mA g^{-1}).

13.7 XPS Valence Band and Segregation Effect in Nanocrystalline Mg$_2$Ni Materials

The valence band and surface segregation effect of microcrystalline and nanocrystalline Mg$_2$Ni alloys and Mg$_2$Ni/Pd composites using X-ray photoelectron spectroscopy (XPS) with depth profile analysis were studied by Smardz et al. [18, 20, 21, 35, 42, 45]. The MA process has been studied by XRD investigations, using Mg$_2$Ni as representative alloy example. Figure 13.18a and 18b show XRD patterns of an initial state (elemental powder mixture) of Mg and Ni powders, respectively. Formation of the nanocrystalline alloy was achieved by annealing the amorphous material in high-purity argon atmosphere at 450°C for 1 h (Fig. 13.18d). All diffraction peaks were assigned to those of the hexagonal crystal structure with cell parameters a = 5.216 Å, c = 13.246 Å. The average size of nanocrystalline grains, according to AFM studies, was of the order of 30 nm. Curve (e) in Fig. 13.18 represents XRD spectrum of the Mg$_2$Ni/Pd composite prepared by MA for 1 h of nanocrystalline Mg$_2$Ni (see Fig. 13.18d) mixed with 10 wt% Pd powder.

Figure 13.19a shows XRD pattern of mixture of 2Mg and Ni powders after MA for 48 h followed by annealing at 450°C for 1 h. XRD results measured for pure powder of palladium are presented in Fig. 13.19b. Curve (c) in Fig. 13.19 represents XRD spectrum of Mg$_2$Ni/Pd (I) composite prepared by HEBM for 1 h of amorphous Mg$_2$Ni (see Fig. 13.19c) mixed with 10 wt% Pd powder. In Fig. 13.19d XRD spectrum measured for Mg$_2$Ni/Pd (II) composite prepared by MA for 48 h followed by annealing at 450°C for 1 h of 2Mg and Ni powders mixed with 10 wt% Pd powder is shown.

The different intensities and positions of the XRD peaks shown in Fig. 13.19c,d could be explained by different grain sizes and distribution of Pd atoms in Mg$_2$Ni/Pd (I) and (II) composites. The above behavior could be better visible in the scanning electron microscopy images shown in Fig. 13.20 and 13.21 for the Mg$_2$Ni/Pd (I) and (II) composites, respectively. Especially, we have observed non-uniform distribution of Pd atoms in the Mg$_2$Ni/Pd (I) composite compared to that observed for the Mg$_2$Ni/Pd (II) composite. On the other hand, distributions of Mg and Ni atoms are rather uniform in the both composites (see Figs. 13.20 and 13.21).

As it was mentioned earlier, the poor cyclic behavior of Mg$_2$Ni electrodes is attributed to the formation of Mg(OH)$_2$ on the electrodes, which has been considered to arise from the charge-discharge cycles. To avoid the surface oxidation of Mg$_2$Ni-type material the effect of palladium coating was studied. The discharge capacities up to 305 and 205 mA h g^{-1} at the first cycle were measured for the Mg$_2$Ni/Pd (I) and (II) composites (see Fig. 13.19), respectively. Mechanically coated Mg$_2$Ni material with palladium has effectively reduced the degradation rate of the studied electrode materials. It is important to note that discharge capacities measured for the Mg$_2$Ni/Pd (I) degraded strongly with cycling in comparison to Mg$_2$Ni/Pd (II) composite.

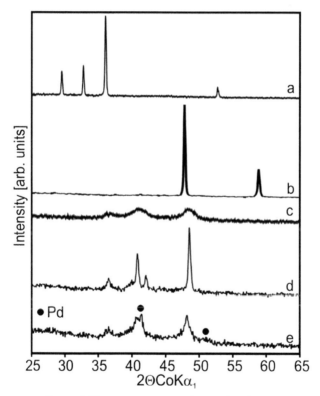

Figure 13.18 X-ray diffraction patterns (Co-K$_{\alpha 1}$) of pure powders of magnesium (a) and nickel (b), and mixture of 2Mg and Ni powders after MA for 45 h (c) followed by annealing at 450°C for 1 h (d). Curve (e) represents XRD spectrum of Mg$_2$Ni/Pd composite prepared by MA for 1 h of nanocrystalline Mg$_2$Ni (see part (d)) mixed with 10 wt% Pd powder.

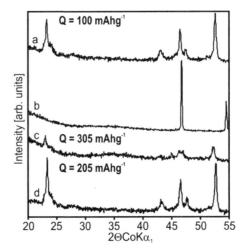

Figure 13.19 X-ray diffraction patterns (Co-K$_{\alpha 1}$) of mixture of 2Mg and Ni powders after MA for 48 h followed by annealing at 450°C for 1 h (a) and pure powder of palladium (b). Curve (c) represents XRD spectrum of Mg$_2$Ni/Pd (I) composite prepared by HEBM for 1 h of amorphous Mg$_2$Ni (see Fig. 13.1c) mixed with 10 wt% Pd powder. XRD spectrum (d) was measured for Mg$_2$Ni/Pd (II) composite prepared by MA for 48 h followed by annealing at 450°C for 1 h of 2Mg and Ni powders mixed with 10 wt% Pd powder.

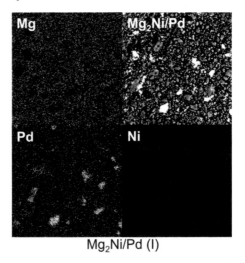

Figure 13.20 Scanning electron microscopy image of Mg$_2$Ni/Pd (I) composite prepared by HEBM for 1 h of amorphous Mg$_2$Ni mixed with 10 wt% Pd powder. Distributions of Mg, Ni, and Pd atoms are also shown. Magnification: 1500×.

XPS Valence Band and Segregation Effect in Nanocrystalline Mg$_2$Ni Materials | 249

Mg$_2$Ni/Pd (II)

Figure 13.21 Scanning electron microscopy image of Mg$_2$Ni/Pd (II) composite prepared by MA for 48 h followed by annealing at 450°C for 1 h of 2Mg and Ni powders mixed with 10 wt% Pd powder. Distributions of Mg, Ni, and Pd atoms are also shown. Magnification: 1500×.

The experimental XPS valence bands measured for the nanocrystalline MA Mg$_2$Ni alloy and microcrystalline Mg$_2$Ni thin film are shown in Fig. 13.22. The experimental XPS valence bands measured for MA nanocrystalline alloy showed a significant broadening compared to that obtained for the microcrystalline thin film. Furthermore a significant band broadening was also observed for the Mg$_2$Ni/Pd composite prepared by MA for 1 h of nanocrystalline Mg$_2$Ni mixed with 10 wt% Pd powder (see curve (a) in Fig. 13.23). For the "as prepared" composite two peaks near 3.5 and 7 eV binding energy (see curve (a) in Fig. 13.23) could be observed. These peaks disappear after UHV annealing at 450°C for 1 h. This behavior could be explained mainly by non-uniform distribution of Pd in the MA nanocrystalline Mg$_2$Ni material. After UHV annealing, the distribution of Pd in the nanocrystalline Mg$_2$Ni became more uniform and the two peaks observed in Fig. 13.23 (curve (a)) disappear. This is also consistent with the scanning electron microscopy studies for the Mg$_2$Ni/Pd (I) and (II) composites shown in Fig. 13.20 and 13.21, respectively.

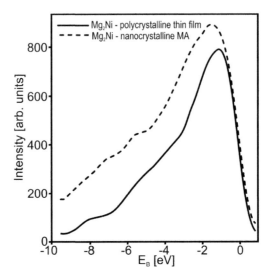

Figure 13.22 XPS valence band (Al-K$_\alpha$) spectra for MA nanocrystalline Mg$_2$Ni alloy and microcrystalline Mg$_2$Ni thin films The XPS measurements were performed immediately after heating in UHV conditions followed by removing of a native oxide and possible impurities layer using ion gun etching system (see text).

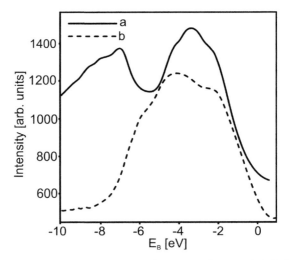

Figure 13.23 XPS valence band (Al-K$_\alpha$) spectra for Mg$_2$Ni/Pd composite prepared by MA for 1 h of nanocrystalline Mg$_2$Ni with 10 wt% Pd powder (curve a) and after UHV annealing at 450°C for 1 h (curve b). The XPS measurements were performed immediately after removing of a native oxide and possible impurities layer using ion gun etching system.

Results on XRF measurements revealed the assumed bulk chemical composition of the microcrystalline and nanocrystalline Mg$_2$Ni-type alloys. On the other hand, core–level XPS showed that the surface segregation of Mg atoms in the MA nanocrystalline samples is stronger compared to that of microcrystalline thin films. Especially, a strong surface segregation of Mg atoms we have observed for the Mg$_2$Ni/Pd composites. Figure 13.24 shows normalized integral intensities of Mg, O, Ni, and Pd XPS peaks versus sputtering time as converted to depth for Mg$_2$Ni/Pd composite. The XPS Mg-1s, Ni-2p$_{3/2}$, and Pd-3d$_{5/2}$ peaks were normalized to the intensities of in-situ prepared pure Mg, Ni, and Pd thin films, respectively. The oxygen 1s peak was normalized to the O-1s intensity in the MgO single crystal. The results presented in Fig. 13.24 show that Ni and Pd atoms are practically absent on the composite surface. On the other hand, Mg atoms strongly segregate to the surface and form a Mg based oxide layer under atmospheric conditions.

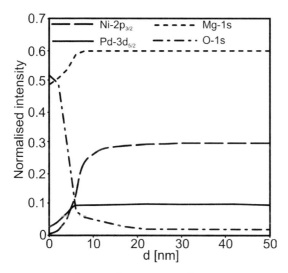

Figure 13.24 Normalized integral intensities of Mg, O, Ni, and Pd XPS peak versus sputtering time as converted to depth for Mg$_2$Ni/Pd composite. The XPS Mg-1s, Ni-2p$_{3/2}$, and Pd-3d$_{5/2}$ peaks were normalized to the intensities of in-situ prepared pure Mg, Ni, and Pd thin films, respectively. The oxygen 1s peak was normalized to the O-1s intensity in the MgO single crystal. The XPS measurements were performed immediately after heating in UHV conditions (see text) which allow the removal of adsorbed impurities (mainly carbonates) excluding a stabile oxide top layer.

The oxidation process is depth limited such that an oxide-covering layer with a well-defined thickness is formed by which the lower lying metal is prevented from further oxidation. In this way one can obtain a self-stabilized oxide–metal structure. The lower lying Ni and Pd atoms form a metallic subsurface layer and are responsible for the observed relatively high hydrogenation rate [46, 47]. The surface segregation process of Mg atoms in Mg_2Ni/Pd composite is stronger compared to that observed for the Mg_2Ni nanocrystalline alloy. Furthermore, we have observed no segregation effect for the in-situ prepared microcrystalline Mg_2Ni thin films. On the other hand, the Mg_2Ni thin films naturally oxidized in air for 24 h show a small segregation effect of the Mg atoms to the surface.

13.8 Mg-Based Nanocomposite Hydrides for Room Temperature Storage

The influence of microstructure on the structural and thermodynamic properties on $Mg_{1.5}Mn_{0.5}Ni/LaNi_{3.75}Mn_{0.75}Al_{0.25}Co_{0.25}$ and $Mg_{1.5}Mn_{0.5}Ni/Zr_{0.35}Ti_{0.65}V_{0.85}Cr_{0.26}Ni_{1.30}$ hydride materials was studied (Fig. 13.25) [48]. The results show that nanostructured 50% $Mg_{1.5}Mn_{0.5}Ni$ + 50% $LaNi_{3.75}Mn_{0.75}Al_{0.25}Co_{0.25}$ and 75% $Mg_{1.5}Mn_{0.5}Ni$ + 25% $Zr_{0.35}Ti_{0.65}V_{0.85}Cr_{0.26}Ni_{1.30}$ composite materials releases 1.65 and 1.38 wt% hydrogen at room temperature, respectively (Figs. 13.26A,B). This is higher than in nanocrystalline $Mg_{1.5}Mn_{0.5}Ni$ (0.68 wt% at 300°C), $LaNi_{3.75}Mn_{0.75}Al_{0.25}Co_{0.25}$ (1.03 wt%), $Zr_{0.35}Ti_{0.65}V_{0.85}Cr_{0.26}Ni_{1.30}$ (1.31 wt%). It is to be concluded that the observed effect of $LaNi_{3.75}Mn_{0.75}Al_{0.25}Co_{0.25}$ or $Zr_{0.35}Ti_{0.65}V_{0.85}Cr_{0.26}Ni_{1.30}$ nanopowders on hydrogen storage behavior of $Mg_{1.5}Mn_{0.5}Ni$ phase strongly suggest that all the improvement of hydrogen storage properties is due to kinetic effect of $LaNi_5$ or ZrV_2 nanopowder acting as a powerful catalytic rather than to a modification of thermodynamic properties of the $Mg_{1.5}Mn_{0.5}Ni$ + $LaNi_{3.75}Mn_{0.75}Al_{0.25}Co_{0.25}$ or $Mg_{1.5}Mn_{0.5}Ni$ + $Zr_{0.35}Ti_{0.65}V_{0.85}Cr_{0.26}Ni_{1.30}$ systems.

The effects of TiNi alloy addition on the electrochemical and thermodynamic properties of Mg_2Ni at room temperature were also investigated. The nanoscale $Mg_{1.5}Mn_{0.5}Ni$- and $TiFe_{0.25}Ni_{0.75}$-type alloys as well as $Mg_{1.5}Mn_{0.5}Ni/TiFe_{0.25}Ni_{0.75}$ hydride nanocomposites were synthesized (Fig. 13.27, Table 13.4).

Mg-Based Nanocomposite Hydrides for Room Temperature Storage | 253

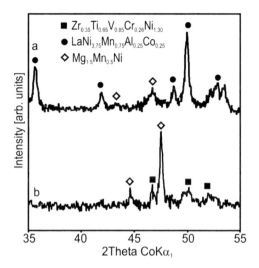

Figure 13.25 XRD spectra of 50% $Mg_{1.5}Mn_{0.5}Ni$/50% $LaNi_{3.75}Mn_{0.75}Al_{0.25}Co_{0.25}$ (a) and 75% $Mg_{1.5}Mn_{0.5}Ni$/25% $Zr_{0.35}Ti_{0.65}V_{0.85}Cr_{0.26}Ni_{1.30}$ (b) nanocomposites after 1 h ball milling.

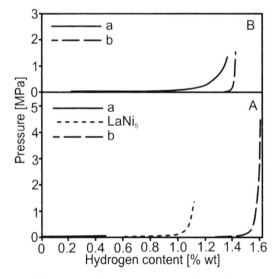

Figure 13.26 (A) PC isotherms at 21°C for (a) composite 50% Mg_2Ni/50% $LaNi_5$ (no hydrogen desorption) and (b) 50%$Mg_{1.5}Mn_{0.5}Ni$/50%$LaNi_{3.75}Mn_{0.75}Al_{0.25}Co_{0.25}$ (samples a and b: composites after nanocomposites after 1 h ball milling. (B) PC isotherms at 21°C for: (a) microcrystalline $Zr_{0.35}Ti_{0.65}V_{0.85}Cr_{0.26}Ni_{1.30}$ alloy and (b) nanocomposite 75% $Mg_{1.5}Mn_{0.5}Ni$/25% $Zr_{0.35}Ti_{0.65}V_{0.85}Cr_{0.26}Ni_{1.30}$ after 1 h ball milling.

Table 13.4 Selected properties of studied nanocrystalline and nanocomposite materials

Composition	Hydrogen content at RT [wt%]	Max. discharge capacity [mAh/g]	Capacity retaining rate after 60th cycle [%]*
$Mg_{1.5}Mn_{0.5}Ni$	0	141	7
$TiFe_{0.25}Ni_{0.75}$	1.30	93	22
75% $Mg_{1.5}Mn_{0.5}Ni$ − 25%$TiFe_{0.25}Ni_{0.75}$	1.40	134	79

*Capacity retaining rate, which is defined as $C_n/C_{max} \times 100\%$. C_{max} is the maximum discharge capacity; C_n is the discharge capacity at nth charge/discharge cycle step.

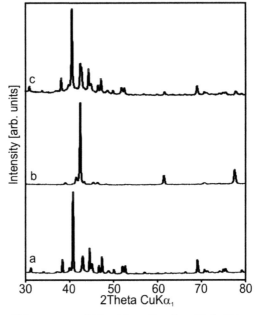

Figure 13.27 XRD spectra of $Mg_{1.5}Mn_{0.50}Ni$ after 72 h MA and annealing 723 K/1 h (a), $TiFe_{0.25}Ni_{0.75}$ after 10 h MA and annealing 973 K/0.5 h (b) and nanocomposite 75 wt% $Mg_{1.5}Mn_{0.5}Ni$/25 wt% $TiFe_{0.25}Ni_{0.75}$ after 1 h ball milling (c).

As was mentioned earlier, the Mg_2Ni electrode, mechanically alloyed and annealed, displayed the maximum discharge capacity at the first cycle but degraded strongly with cycling. The poor cyclic behavior of Mg_2Ni electrodes is attributed to the formation of Mg(OH)$_2$ on the electrodes, which has been considered to arise from the charge-discharge cycles. To avoid the surface oxidation, we have examined the effect of magnesium substitution by Mn in Mg$_2$Ni-type material. This alloying greatly improved the discharge capacities.

The influence of microstructure on the structural, electrochemical and thermodynamic properties of Mg$_{1.5}$Mn$_{0.5}$Ni/ TiFe$_{0.25}$Ni$_{0.75}$ hydride materials was studied. Average particle size values equal 163 and 138 nm for Mg$_{1.5}$Mn$_{0.5}$Ni and TiFe$_{0.25}$Ni$_{0.75}$, respectively. Discharge capacities as a function of cycle number of electrodes prepared with all studied alloys and composites are presented on Fig. 13.28. Electrodes displayed maximum capacities at second cycle. The best activation properties have Mg-based material with TiFe$_{0.25}$Ni$_{0.75}$ phase.

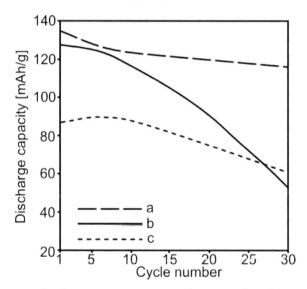

Figure 13.28 Discharge capacities as a function of cycle number of 75 wt% Mg$_{1.5}$Mn$_{0.5}$Ni/25 wt% TiFe$_{0.25}$Ni$_{0.75}$ (a), Mg$_{1.5}$Mn$_{0.50}$Ni (b), TiFe$_{0.25}$Ni$_{0.75}$ (c).

The addition of TiNi(Fe) phase could result in an increase in specific surface area of alloy crystallites and decrease in activation energy barrier of phase transformation. Bigger surface area is mostly caused by pulverization process. The results show that nanostructured 75% Mg$_{1.5}$Mn$_{0.5}$Ni – 25% TiFe$_{0.25}$Ni$_{0.75}$ composite materials releases 1.40 wt% of hydrogen at room temperature (Fig. 13.29). This is higher than in nanocrystalline $Mg_{1.5}Mn_{0.5}Ni$ (0 wt% at RT) and $TiFe_{0.25}Ni_{0.75}$ (1.30 wt% at RT). It can be concluded that the observed effect of $TiFe_{0.25}Ni_{0.75}$ nanopowders on hydrogen storage behavior of $Mg_{1.5}Mn_{0.5}Ni$ phase strongly suggest that all the improvement of hydrogen storage properties is due to kinetic effect of $TiFe_{0.25}Ni_{0.75}$ nanopowder acting as a powerful catalytic rather than to a modification of thermodynamic properties of the $Mg_{1.5}Mn_{0.5}Ni$–$TiFe_{0.25}Ni_{0.75}$ system.

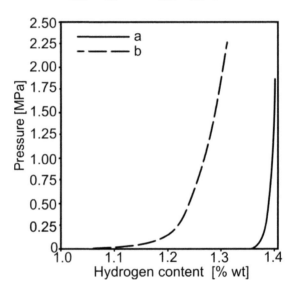

Figure 13.29 Pressure-composition isotherms at 21°C: 75 wt% $Mg_{1.5}Mn_{0.5}Ni$/25 wt% $TiFe_{0.25}Ni_{0.75}$ (a), $TiFe_{0.25}Ni_{0.75}$ (b).

Adding $LaNi_5$, ZrV_2 or TiFe in magnesium-based alloy improves the hydrogen storage performance at room temperatures. The results show that nanostructured composite 50% La(Ni, Mn, Al, Co)$_5$/50% (Mg,Mn)$_2$Ni, 25% (Zr,Ti) (V,Cr,Ni)$_{2.4}$/75% (Mg,Mn)$_2$Ni, and 25% $TiFe_{0.25}Ni_{0.75}$/75% $Mg_{1.5}Mn_{0.5}Ni$ composite materials releases 1.65, 1.38, and 1.40 wt% hydrogen at 21°C, respectively. The nanocrystalline metal hydrides offer a break through in prospects for practical applications of Mg_2Ni phase in hydrogen economy.

References

1. Reilly, J. J., and Wiswall, R. H. (1968). The reaction of hydrogen with alloys of magnesium and nickel and the formation of Mg$_2$NiH$_4$. *Inorg. Chem.* **7**: 2254–2256.
2. Aymard, L., Ichitsubo, M., Uchida, K., Sekreta, E., and Ikazaki, F. (1997). Preparation of Mg$_2$Ni base alloy by the combination of mechanical alloying and heat treatment at low temperature. *J Alloys Comp.* **259**: L5–L8.
3. Imamura, H., Masanari, K., Kusuhara, M., Katsumoto, H., Sumi, T., Sakata, Y. (2005). High hydrogen storage capacity of nanosized magnesium synthesized by high energy ball-milling. *J. Alloys Comp.* **386**: 211–216.
4. Gasiorowski, A., Iwasieczko, W., Skoryna, D., Drulis, H., Jurczyk, M. (2004). Hydriding properties of nanocrystalline Mg$_{2-x}$M$_x$ Ni alloys synthesized by mechanical alloying (M = Mn, Al). *J. Alloys Comp.* **364**: 283–288.
5. Varin, R. A., Czujko, T., Wronski, Z. S. (2009). *Nanomaterials for Solid State Hydrogen Storage*, Springer Science+Business Media, LLC.
6. Anani, A., Visintin, A., Petrov, K., Srinivasan, S. (1994). Alloys for hydrogen storage in nickel/hydrogen and nickel/metal hydride batteries. *J. Power Sources* **47**: 261–275.
7. Zaluski, L., Zaluska, A., and Ström-Olsen, J. O. (1995). Hydrogen absorption in nanocrystalline Mg$_2$Ni formed by mechanical alloying. *J. Alloys Comp.* **217**: 245–249.
8. Orimo, S. I., and Fujii, H. (1998). Effects of nanometer-scale structure on hydriding properties of Mg–Ni alloys: A review. *Intermetallics* **6**: 185–192.
9. Jurczyk, M. (2004). The progress of nanocrystalline hydride electrode materials, *Bull. Pol. Acc., Tech.* **52**: 67–77.
10. Au, M., Pourarian, F., Simizu, S., Sankar, S. G., and Zhang, L. (1995). Electrochemical properties of TiMn$_2$-type alloys ball-milled with nickel powder. *J. Alloys Comp.* **223**: 1–5.
11. Sun, D., Latroche, M., and Percheron-Guégan, A. (1997). Activation behaviour of mechanically Ni-coated Zr-based laves phase hydride electrode. *J. Alloys Comp.* **257**: 302–305.
12. Iwakura, C., Inoue, H., Nohara, S., Shin-ya, R., Kurosaka, S., and Miyanohara, K. (2002). Effects of surface and bulk modifications

on electrochemical and physicochemical characteristics of MgNi alloys. *J. Alloys Comp.* **330–332**: 636–639.

13. Bouaricha, S., Dodelet, J. P., Guay, D., Huot, J., and Schultz, R. (2001). Activation characteristics of graphite modified hydrogen absorbing materials. *J. Alloys Comp.* **325**: 245–251.

14. David, E. (2005). An overview of advanced materials for hydrogen storage. *J. Mater. Process. Technol.* **162–163**: 169–177.

15. Jurczyk, M., Okonska, I., Iwasieczko, W., Jankowska, E., and Drulis, H. (2007). Thermodynamic and electrochemical properties of nanocrystalline Mg_2Cu-type hydrogen storage materials. *J. Alloys Comp.* **429**: 316–320.

16. Okonska, I., Iwasieczko, W., and Jurczyk, M. (2007). Hydriding properties of nanocrystalline Mg_2Cu alloy synthesized by mechanical alloying. *Acta Metal. Slov.* **13**: 70–74.

17. Okonska, I., Iwasieczko, W., Jarzebski, M., Nowak, M., and Jurczyk, M. (2007). Hydrogenation properties of amorphous 2Mg+Fe/x wt% Ni materials prepared by mechanical alloying (x = 0, 100, 200). *Int. J. Hydrogen Energy* **32**: 4186–4190.

18. Jurczyk, M., Smardz, L., Okonska, I., Jankowska, E., Nowak, M., and Smardz, L. (2008). Nanoscale Mg-based materials for hydrogen storage. *Int. J. Hydrogen Energy* **33**, pp. 374–380.

19. Okonska, I., Nowak, M., Jankowska, E., and Jurczyk, M. (2008). Hydrogen storage by Mg-based nanomaterials. *Rev. Adv. Mater. Sci.* **18**: 627–631.

20. Smardz, L., Jurczyk, M., Smardz, K., Nowak, M., Makowiecka, M., and Okonska, I. (2008). Electronic structure of nanocrystalline and polycrystalline hydrogen storage materials. *Renew. Energy* **33**: 201–210.

21. Smardz, K., Smardz, L., Okonska, I., Nowak, M., and Jurczyk, M. (2008). XPS valence band and segregation effect in nanocrystalline Mg_2Ni-type materials. *Int. J. Hydrogen Energy* **33**: 387–392.

22. Nayeb-Hashemi, A. A., and Clark, J. B. (1985). The Mg–Ni system. *Bull. Alloy Phase Diagrams* **6**: 238–244.

23. Nayeb-Hashemi, A. A., and Clark, J. B. (1984). The Cu-Mg system. *Bull. Alloy Phase Diagrams* **5**: 36–43.

24. Gasiorowski, A. (2004). Influence of microstructure and chemical composition on the properties of Mg-based alloys reversibly absorbing hydrogen. PhD thesis. Poznan University of Technology.

25. Reilly, J. J., and Wismall, R. H. (1968). Metal hydrides for energy storage. *Inorg. Chem.* **7**: 2254–2256.
26. Bradhurst, D. H. (1983). Metal hydrides for energy storage. *Metals Forum* **6**: 139–148.
27. Anani, A., Visintin, A., Srinivasan, S., Appleby, A. J., Reilly, J. J., and Johnson, J. R. (1992). in Corrigan, D. A., ed., *Proceedings of the Symposium on Hydrogen Storage Materials, Batteries and Electrochemistry*, vol. 5, The Electrochemical Society, Pennington, NJ, USA, p. 105.
28. Mu, D., Hatano, Y., Abe, T., Watanabe, K. J. (2002). Degradation kinetics of discharge capacity for amorphous Mg–Ni electrode. *J. Alloys Comp.* **334**: 232–237.
29. Zhang, S. G., Yorimitsu, K., Nohara, S., Morikawa, T., Inoue, H., Iwakura, C. (1998). Surface analysis of an amorphous MgNi alloy prepared by mechanical alloying for use in nickel–metal hydride batteries. *J. Alloys Comp.* **270**: 123–126.
30. Jurczyk, M., Nowak, M. (2008). Nanomaterials for hydrogen storage synthesized by mechanical alloying, in: Ali, E., ed., *Nanostructured Materials in Electrochemistry*, Wiley, Chapter 9.
31. Stefanov, P. (1996). XPS study of surface segregation in Ni-Mg alloy films, *Vacuum* **47**: 1107–1110.
32. Wang, L. B., Tang, Y. H., Wang, Y. J., Li, Q. D., Song, H. N., and Yang, H. B. (2002). The hydrogenation properties of $Mg_{1.8}Ag_{0.2}Ni$ alloy. *J. Alloys Comp.* **336**: 297–300.
33. Zhang, S. G., Yorimitsu, K., Nohara, S., Morikawa, T., Inoue, H., and Iwakura, C. (1998). Surface analysis of an amorphous MgNi alloy prepared by mechanical alloying for use in nickel-metal hydride batteries. *J. Alloys Comp.* **270**: 123–126.
34. Wang, P., Wang, A. M., Ding, B. Z., and Hu, Z. Q. (2002). Mg–FeTi$_{1.2}$ (amorphous) composite for hydrogen storage. *J. Alloys Comp.* **334**: 243–248.
35. Nowak, M., Okonska, I., Smardz, L., and Jurczyk, M. (2009). Segregation effect on nanoscale Mg-based hydrogen storage materials. *Mater. Sci. Forum* **610–613**: 431–436.
36. Fitzsimmons, M. R., Estman, J. A., Robinson, R. A., Lawson, A. C., Thompson, J. D., and Morshovich, R. (1993). Magnetic order in nanocrystalline Cr and suppression of antiferromagnetism in bcc Cr. *Phys. Rev. B* **48**: 8245–8262.
37. Okonska, I. (2008). Mg-3d (3d = Fe, Co, Ni, Cu) based nanostructured materials reversibly absorbing hydrogen. PhD thesis. Poznan University of Technology.

38. Szajek, A., Jurczyk, M., Smardz, L., Okonska, I., and Jankowska, E. (2007). Electrochemical and electronic properties of nanocrystalline Mg-based hydrogen storage materials. *J. Alloys Comp.* **436**: 345–350.
39. Garcia, G. N., Abriata, J. P., and Sofo, J. O. (1999). Calculation of the electronic and structural properties of cubic Mg_2NiH_4. *Phys. Rev. B* **59**: 11746–11754.
40. Myers, W. R., Wang, L. W., Richardson, T. J., and Rubin, M. D. (2002). Calculation of thermodynamic, electronic, and optical properties of monoclinic Mg_2NiH_4. *J. Appl. Phys.* **91**: 4879–4885.
41. Iwakura, C., Inoue, H., Zhang, S. G., and Nohara, S. (1999). A new electrode material for nickel-metal hydride batteries: MgNi–graphite composites prepared by ball-milling. *J. Alloys Comp.* **293–295**: 653–657.
42. Smardz, L., Smardz, K., Nowak, M., and Jurczyk, M. (2001). Structure and electronic properties of $La(Ni,Al)_5$ alloys. *Cryst. Res. Tech.* **36**: 1385–1392.
43. Nowak, M., Smardz, L., and Jurczyk, M. (2010). Hydrogen storage in nanostructured Mg-based hydrides and their composites. *Curr. Topics Electrochem.* **15**: 25–38.
44. Xiao, X. H., Wang, X. H., Gao, L. H., Wang, L., and Chen, C. G. (2006). Electrochemical properties of amorphous MgFe alloys mixed with Ni prepared by ball milling. *J. Alloys Comp.* **413**: 312–318.
45. Szajek, A., Okonska, I., and, Jurczyk, M. (2007). Electronic and electrochemical properties of Mg_2Ni alloy doped by Pd atoms. *Mater. Sci. Poland* **25**: 1251–1257.
46. Siegmann, H. C., Schlapbach, L., and Brundle, C. R. (1978). Self-restoring of the active surface in the hydrogen sponge $LaNi_5$. *Phys. Rev. Lett.* **40**: 972–976.
47. Schlapbach, L. (1992). Surface properties and activation, hydrogen in intermetallic compounds, in Schlapbach, L., ed., *Hydrogen in Intermetallic Compounds II*.
48. Jurczyk, M., Nowak, M., Smardz, L., and Szajek, A. (2011). Mg-based nanocomposites for room temperature hydrogen storage, *TMS Ann. Meeting* **1**: 229–236.

Chapter 14

(La, Mg)$_2$Ni$_7$-Based Hydrogen Storage Alloys

Marek Nowak and Mieczyslaw Jurczyk

*Institute of Materials Science and Engineering,
Poznan University of Technology, Poznan, Poland*

marek.nowak@put.poznan.pl, mieczysław.jurczyk@put.poznan.pl

14.1 Phase Diagram and Structure

Among many materials, magnesium-based materials have been intensively investigated as promising candidates for hydrogen storage due to their high hydrogen storage capacity (7.6 wt%), low cost and light weight [1–4]. However, the application of Mg-based materials is hindered by their slow kinetics and relatively high operating temperature [5]. In order to overcome these drawbacks, the addition of certain rare earth (RE) metals [6–26] and alloying some intermetallic compounds with Mg-based alloys have been considered by many researchers.

As a result, RE–Mg–Ni-based hydrogen storage alloys currently have attracted an increasing amount of attention as next generation negative electrode materials for Ni-MH$_x$ batteries with high energy density. For practical applications, numerous

---

Handbook of Nanomaterials for Hydrogen Storage
Edited by Mieczyslaw Jurczyk
Copyright © 2018 Pan Stanford Publishing Pte. Ltd.
ISBN 978-981-4745-66-6 (Hardcover), 978-1-315-36444-5 (eBook)
www.panstanford.com

efforts to improve the overall electrochemical properties of RE–Mg–Ni-based electrode alloys have been undertaken and studies of interest include optimization of composition, high-energy ball-milling treatment, heat treatment, surface treatment, electrolyte modification, operating temperatures and degradation mechanism.

High hydrogen capacity, moderate hydrogen equilibrium pressure, and light and less expensive elements make them remarkable from an economical point of view. Recently, details about the structures of La_2MgNi_9, $La_5Mg_2Ni_{23}$, La_3MgNi_{14} [7], La_2MgNi_9, $La_5Mg_2Ni_{23}$, La_3MgNi_{14} [8], La–Mg–Ni_x (x = 3–4) [9], $La_{0.67}Mg_{0.33}Ni_{2.5}Co_{0.5}$ [12], $La_{2-x}Mg_xNi_{7.0}$ (x = 0.3–0.6) [14], $La_{2-x}Mg_xNi_{7.0}$ (x = 0.3–0.6) [15] have been published. Generally, the hydrogen storage properties of ternary RE–Mg–Ni compounds are superior to corresponding binary AB_n (n = 2–5) compounds [6–26].

The introduction of Mg into AB_{2-5}-type rare earth–based hydrogen storage alloys facilitates the formation of a $(La,Mg)Ni_3$ phase with a rhombohedral $PuNi_3$-type structure or a $(La,Mg)_2Ni_7$ phase with a hexagonal Ce_2Ni_7-type structure (Fig. 14.1, Table 14.1). These phases possess long-periodic one-dimensional superstructures in which the AB_5 unit ($CaCu_5$-type structure) and the AB_2 unit (Laves structure) are rhombohedrally or hexagonally stacked with a ratio of n:1 along the c-axis direction, consequently resulting in a larger hydrogen storage capacity. The presence of Mg in $LaNi_3$ phase and in La_2Ni_7 phase prevents the amorphization of their hydride phases, leading to a higher electrochemical capacity.

Table 14.1 Phases in the La–Mg–Ni system

Phase	Space group
AB_2	$F\bar{4}3m$
AB_3	$R\bar{3}m$, $P6_3$, mmc
A_2B_7	$R\bar{3}m$, $P6_3$, mmc
A_5B_{19}	$R\bar{3}m$, $P6_3$, mmc
A_6B_{24}	$R\bar{3}m$, $P6_3$, mmc
AB_5	$P6/mmm$

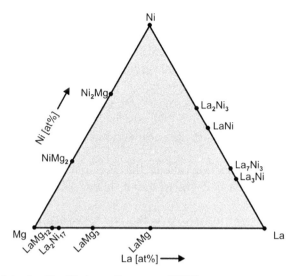

Figure 14.1 La–Mg–Ni phase diagram at 500°C.

14.2 Electrochemical Properties

Electrode performance was found to depend strongly on the stoichiometric ratio, alloy components and microstructure. The $AB_{3.0}$- and $AB_{3.5}$-type RE–Mg–Ni-based alloys exhibit a higher discharge capacity and better electrochemical kinetics. The specific function of the constituent elements including Mg, Ce, Pr, Nd, Co, Mn, Al, Zr, W, Fe, Cu, and Cr are discussed. Optimum compositions contain mainly metallic elements of La, Mg, Ni, Co, Mn, and Al. Annealing treatment significantly increases the discharge capacity, improves the cyclic stability and enhances the high rate dischargeability. The optimized annealing temperature lies between 850 and 950°C. Mechanistic investigations indicate that the pulverization of alloy particles and the oxidation/corrosion of active components during cycling are the two main factors responsible for the fast capacity degradation of RE–Mg–Ni-based alloy electrodes, and the degradation progress is divided into three consecutive stages including pulverization and La oxidation, and oxidation-passivation. Consequently, a decrease in the pulverization of the alloy particles and an increase in their anti-oxidation/corrosion ability are necessary for an improvement in cyclic stability. Co and Al produce these effects in RE–Mg–Ni-based electrode alloys. Despite the overall electrochemical properties of RE–Mg–Ni-based electrode alloys being significantly improved over the last decade, their practical application is still a challenge because of the industrialized preparation techniques and the cost of alloys. Therefore, optimized preparation techniques for industrial scale and low-cost RE–Mg–Ni-based electrode alloys (low-Co or Co-free) with a high discharge capacity, a long cycle life and good kinetics are still needed to satisfy the requirements for Ni-MH$_x$ batteries.

Detailed information about the function of the individual elements used in RE–Mg–Ni-based alloys is summarized in Table 14.2.

It is important to note, that the crystal structure of a binary RE$_2$Ni$_7$ compound is size dependent. The Ce$_2$Ni$_7$-type structure is stable for larger RE-atomic radii and the Gd$_2$Co$_7$-type structure is preferred for smaller M-atomic radii. Both structures coexist in the case of medium-sized RE-atomic radii [13].

The structure stabilities of Ce$_2$Ni$_7$- or Gd$_2$Co$_7$-type in multicomponent (RE,Mg)$_2$Ni$_7$ compounds were studied by comparing their relative amounts in the (La$_{1.66}$Mg$_{0.34}$)Ni$_7$-based alloy after partial substitution by different elements. On the basis of the (La$_{1.66}$Mg$_{0.34}$)Ni$_7$ compound, Ce, Pr, Nd, Y, Sm, and Gd are used as smaller substitutes for La, respectively, to change the average A-atomic radius. Additionally, Ni is partially replaced by Co and Al to increase the average B-atomic radius. Nickel has the smallest atomic radius among all transition metals.

The (La$_{1.66}$Mg$_{0.34}$)Ni$_7$ alloy with a Ce$_2$Ni$_7$-type structure can absorb and desorb hydrogen under moderate conditions [14, 15]. Its hydride formation enthalpy is about −31.4 kJ/mol H$_2$, which is close to −30 kJ/mol H$_2$ for the LaNi$_5$–H$_2$ system. However, electrochemical properties of (La$_{1.66}$Mg$_{0.34}$)Ni$_7$-based compounds have not been reported to date. Therefore, the effects of partial substitution for La by Mg in the (La$_{1.66}$Mg$_{0.34}$)Ni$_7$ compound on the electrochemical hydrogen absorption and desorption is interesting from the application point of view of these materials in hydrogen storage. The aim of our research is to develop a new generation of nanostructured (La, Mg)$_2$Ni$_7$ alloys for hydrogen energy infrastructure building and study of influence of their microstructure and chemical composition on electrochemical storage properties.

Table 14.2 The influence of selected elements for the properties of RE–Mg–Ni-based alloys

Element	Properties
La	Increase the unit cell volume Improve the plateau properties Increase the discharge capacity, easy activation and good high-rate dischargeability Poor cyclic stability due to the corrosion of La and large unit cell expansion rate
Mg	Eliminate the amorphization of hydrides Decrease the unit cell volume and the stability of hydride Increase the discharge capacity, the high-rate dischargeability and the cyclic stability
Ni	The smallest atomic radius among all transition metals Forms unstable hydride ($\Delta H = -3$ kJ/mol for $NiH_{0.5}$) Indispensable element because of its high electrocatalytic activity Forms intermetallics, decreases the Me–H bond strength to a suitable level Sensitive to corrosion and oxidation during cycling—forms $Ni(OH)_2$ Extensively high content leads to decrease in the discharge capacity
Co	Increase the lattice parameters and cell volumes Decrease the plateau pressure and the hysteresis of hydrogen absorption and desorption
	Does not form unstable hydrides ($-\Delta H = +15$ kJ/mol for $CoH_{0.5}$) Increase the hydrogen storage capacity Improve effectively the cyclic stability due to the decrease of cell volume change and the increase of the surface passivation Increase the electrochemical kinetics
Al	Increase the lattice parameters and cell volumes Forms unstable hydride ($\Delta H = -7$ kJ/mol for $AlH_{0.5}$, $\Delta H = -4$ kJ/mol for AlH_3) Decrease the plateau pressure, increase the plateau slope and reduce the hydrogen storage capacity Improve significantly the cyclic stability due to the formation of a dense oxide film Decrease the maximum discharge capacity and the high-rate dischargeability

Microstructural La–Mg–Ni compounds with A_2B_7 type structure are already utilized in novel Ni-MH$_x$ batteries [12–15]. La–Mg–Ni-system A_2B_7-type alloys are considered to be the most promising candidates as the negative electrode materials of Ni-MH$_x$ rechargeable battery in virtue of their higher discharge capacities (above 300 mAh/g) and lower production costs [27–29].

However, La–Mg–Ni (PuNi$_3$-type) alloy electrodes suffered from serious degradation of capacity during charge/discharge cycles. Until now, in order to improve the electrochemical performance of PuNi$_3$-type alloy electrodes, the partial replacement of nickel by transition metals or lanthanum by misch metal has been conducted and studied systematically, but the cyclic stability of alloy electrodes has not been improved effectively [13, 14].

Kohno et al. studied hydrogen storage properties of ternary La$_2$MgNi$_9$, La$_5$Mg$_2$Ni$_{23}$ and La$_3$MgNi$_{14}$ system alloys [8]. Within 30 cycles, the La$_5$Mg$_2$Ni$_{23}$ alloy electrode showed perfect cyclic stability. The discharge capacity for this alloy was 410 mAh/g.

The as cast and annealed La$_{1.5}$Mg$_{0.5}$Ni$_7$ hydrogen storage alloys exhibited a multi-phase microstructure, such as Gd$_2$Co$_7$-, Ce$_2$Ni$_7$-, PuNi$_3$-, CaCu$_5$-, and MgCu$_4$Sn-type phase [27]. Annealing treatment results in the evolution of phase structure from a multi-phase to a double-phase. After annealing treatment at 800°C, the CaCu$_5$- and MgCu$_4$Sn-type phases disappeared rapidly. Additionally, with the increase of annealing temperature the abundance of PuNi$_3$-type phase decreased and disappeared after heat treatment at 900°C. At the same time, this annealing treatment led to the formation of a large amount of Ce$_2$Ni$_7$ phase. Gd$_2$Co$_7$ phase was mainly formed from the cooling process of molten metallic solution.

A combination of the method of preparation and optimization of chemical composition of ternary (RE–Mg)$_2$Ni$_7$-type materials can affect the properties of new generation of metal hydrides. The results obtained for the studies conducted so far in our research group are very promising [27]. (La, Mg)$_2$Ni$_7$ nanomaterials were synthesized by mechanical alloying. The research will allow link process parameters (MA, HEBM) and the selection of the chemical composition to determine their impact on the desired hydrogen sorption properties. Significant improvement of the electrochemical discharge storage capacity can be achieved.

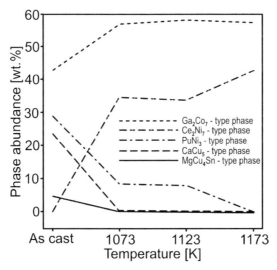

Figure 14.2 Phase abundance as a function of annealing temperature in $La_{1.5}Mg_{0.5}Ni_7$ alloy prepared by induction melting followed y different annealing treatments (800 and 900°C) for 24 h.

14.3 (La,Mg)$_2$Ni$_7$-Type Alloys Synthesized by Mechanical Alloying

Recently, we have synthesized nanoscale (La,Mg)$_2$Ni$_7$-type samples [27]. The behavior of MA process has been studied by X-ray diffraction investigations. Figure 14.3 shows a series of XRD spectra of mechanically alloyed La–Mg–Ni powder mixtures for 48 h. It can be seen that only Ni line is observed in La–Mg–Ni powder mixture. Finally, the obtained powder was heat treated in high purity argon atmosphere at 800°C for 0.5 h. All diffraction peaks were assigned to those of the hexagonal crystal structure; a = 5.06 Å and c = 24.57 Å. The average size of La–Mg–Ni powders, according to AFM studies, was of the order of 30 nm. Additionally, in Fig. 14.4, the X-ray diffraction spectra of La$_2$Ni$_7$ (a), La$_{1.5}$Mg$_{0.5}$Ni$_7$ (b) after MA and heat treatment (800°C/0.5 h) are shown.

The activation capability is indicated by the number of charging-discharging cycles required for attaining the greatest discharge capacity through a charging-discharging cycle at a constant current density of 30 mA/g. Figure 14.5 depicts the cycle number dependence of the discharge capacities of the annealed La–Mg–Ni alloy. It is indicated that the alloy possess a superior activation performance, attaining their maximum discharge capacity within one or two charging-discharging cycles. RE_2Ni_7-type alloys have been considered to be the most promising candidates as the negative electrode materials of Ni-MH$_x$ rechargeable battery in virtue of their high discharge capacities and low production costs.

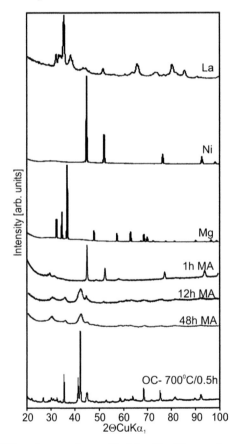

Figure 14.3 XRD patterns of as-milled La–Mg–Ni samples in function of milling time (1 h, 12 h, 48 h) and after annealing at 700°C for 0.5 h [34].

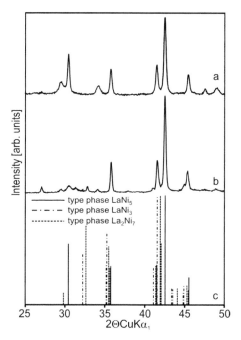

Figure 14.4 X-ray diffraction spectra of La_2Ni_7 (a), $La_{1.5}Mg_{0.5}Ni_7$ (b) after MA and heat treatment (800°C/0.5 h). The positions and intensities of the characteristic lines of $LaNi_3$, $LaNi_5$ and La_2Ni_7 phases are presented (c); (CuKα_1).

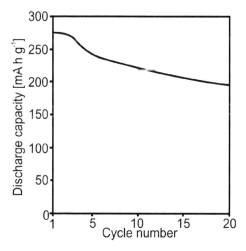

Figure 14.5 Discharge capacity of nanostructured La–Mg–Ni (RE_2Ni_7-type) as a function of cycle number [34].

On Fig. 14.6, the influence of Mg content on the discharge capacity in nanostructured La–Mg–Ni alloy is shown. The highest discharge capacity was observed for $Mg_{0.5}$.

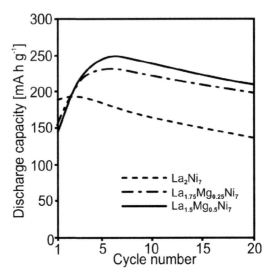

Figure 14.6 Influence of Mg on the discharge capacity of La–Mg–Ni nanostructured alloy.

Recently, the properties of $La_{0.63}RE_{0.2}Mg_{0.17}Ni_{3.3}Co_{0.3}Al_{0.1}$ (RE = La, Ce, Pr, Nd, Sm, Gd, Tb, Dy, Ho, Er, Tm, Yb, Y, Sc) alloys prepared by induction melting followed by annealing treatment at 900°C for 8 h were studied [28, 29]. The alloys consisted mainly of La_2Ni_7 and $LaNi_5$ phases. Substitution of rare earth metals for La was favorable for the formation of a Ce_2Ni_7-type phase.

Electrochemical experiments showed that all alloy electrodes exhibited good activation characteristics, that the discharge capacity improved with the substitution of Ce–Sc for La, and the $La_{0.63}RE_{0.2}Mg_{0.17}Ni_{3.3}Co_{0.3}Al_{0.1}$ alloy possessed the maximum discharge capacity (400.6 mA h g^{-1}) [29]. At the same time, the cyclic stability of alloy electrodes was also improved by substitution of Ce–Sc at La sites, and the capacity retention rate at the 100th cycle was increased by 23.6% for the Gd-substituted alloy. The Yb-substituted alloy possessed excellent high rate dischargeability (HRD900 = 92.84%). All experiments implied that the alloy electrode with RE = Gd possessed excellent overall electrochemical properties.

The positive impact of RE substitution on the cycle stability of the alloy is ascribed to the decrease of the cell parameters caused by such substitution. Additionally, the reasons for the capacity decay of the electrode alloy are the pulverization and oxidation of the alloy during the charging–discharging cycle. The cycle stability of the studied alloy increases with RE substitution. The Ce_2Ni_7-type phase was the main phase of the all alloys.

The experimental results indicate that the cyclic stability of the Ce_2Ni_7-type phase is much better than that of the Gd_2Co_7-type phase. Additionally, rare earth element substitution in $(La,Mg)_2Ni_7$ benefited the formation of a A_2B_7-type phase, the lattice parameters of the Ce_2Ni_7-type phase (in the range of 0.53–0.54 nm^3) influence the electrochemical properties, rare earth element substitution increases the electrochemical discharge capacity and improves cycling stability except for the Yb-substituted alloy. Y substitution exhibits maximum discharge capacity (400.6 mA h g^{-1}) and the capacity retention rate at the 100th cycle of Gd-substituted alloy rises by 23.6%.

In our earlier work, hydrogen storage capacities and the electrochemical discharge capacities of the $Mg_2(Ni,Cu)$-, $LaNi_5$-, ZrV_2-type nanocrystalline alloys and $Mg_2Ni/LaNi_5$-, Mg_2Ni/ZrV_2-type nanocomposites have been measured [30]. The nanocomposite structure reduced hydriding temperature and enhanced hydrogen storage capacity of Mg-based materials. The nanocomposites $(Mg,Mn)_2Ni$ (50 wt%)-$La(Ni,Mn,Al,Co)_5$ (50 wt%) and $(Mg,Mn)_2Ni$ (75 wt%)-$(Zr,Ti)(V,Cr,Ni)_{2.4}$ (25 wt%) materials releases 1.65 wt% and 1.38 wt% hydrogen at 25°C, respectively. The strong modifications of the electronic structure of the nanocrystalline alloys could significantly influence hydrogenation properties of Mg-based nanocomposites.

Apart from nanocrystalline Mg-based materials, mechanically alloyed compounds based on Ti-Ni composition have attracted our interest [31, 32]. TiNi electrode alloy with and without palladium and/or multi-walled carbon nanotubes (MWCNTs) was prepared by ball co-milling. Scanning electron microscopy observations showed that after co-milling with 5 wt% MWCNTs, particles size of TiNi alloy decreased. The TiNi–5 wt% Pd–5 wt% MWCNTs nanocomposite showed the highest discharge capacity (266 mAh/g at the third cycle). Addition of MWCNTs improved the electrode cycle stability (see Chapter 10).

Studied by our research group Pd and MWCNTs can be also applied as catalysts for gaseous hydrogenation process [31, 32]. Pd can improve hydrogenation properties by its good catalytic properties for H_2 dissociation and can reduce the corrosion of materials. On the other hand, MWCNTs provides an easy route for direct transport of desorbed hydrogen. MWCNTs penetrate into thin surface oxide layers, where act as hydrogen pumps reducing the activation energy of hydrogen association on the metal system.

The innovation nature of our research is the combination of method of production (mechanical alloying) with optimization of chemical composition (RE–Mg)$_2$Ni$_7$-type materials which did not take place in the past and is a very promising. As preliminary research shows, this combination can be useful for preparation of hydrogen storage materials [33]. Both mentioned, production method and optimization of chemical composition may significantly change the microstructure, electrochemical and hydrogen sorption properties.

14.4 RE–Mg–Ni-Based Alloy Electrodes

Until now, a lot of optimized RE–Mg–Ni-based electrode alloys have been produced for Ni-MH$_x$ batteries. Some of them are shown in Table 14.3 [34–43].

Table 14.3 Structures and electrochemical properties of some RE–Mg–Ni-based alloy electrodes

Alloy	Structure	Discharge capacity (mAh/g)
La$_2$MgNi$_9$	PuNi$_3$-type	397.5
La$_{0.8}$Mg$_{0.2}$Ni$_{3.75}$	(La,Mg)$_2$Ni$_7$	368.2
Ml$_{0.7}$Mg$_{0.3}$Ni$_{3.2}$	Amorphous	114
La$_{0.7}$Mg$_{0.3}$Ni$_{3.5}$	Ce$_2$Ni$_7$-type, LaNi$_5$, LaMgNi$_4$	352.8
La$_{0.75}$Mg$_{0.25}$Ni$_{3.5}$	(La,Mg)$_2$Ni$_7$, LaNi$_5$	343.7
La$_{1.5}$Mg$_{0.5}$Ni$_7$	(La,Mg)$_2$Ni$_7$, LaNi$_5$	389.48
La$_{1.5}$Mg$_{0.5}$Ni$_{5.2}$Co$_{1.8}$	PuNi$_3$, Ce$_2$Ni$_7$	405.69
Ml$_{0.8}$Mg$_{0.2}$Ni$_{3.2}$Co$_{0.4}$Al$_{0.2}$	LaNi$_5$, LaNi$_3$	327

Alloy	Structure	Discharge capacity (mAh/g)
$La_{0.7}Mg_{0.3}Ni_{2.45}Mn_{0.1}Co_{0.75}Al_{0.2}$	$(La,Mg)Ni_3$, $LaNi_5$	370
$La_{0.4}Nd_{0.4}Mg_{0.2}Ni_{3.2}Co_{0.2}Al_{0.2}$	$LaNi_5$, La_2Ni_7, $LaNi_3$	372
$La_{0.7}Mg_{0.3}Ni_{3.5}$- 40%$Ti_{0.17}Zr_{0.08}V_{0.35}Cr_{0.1}Ni_{0.3}$	$(La,Mg)Ni_3$, $LaNi_5$, LaNi, bcc	335.2

New $(RE-Mg)_2Ni_7$-type nanomaterials can be very important step in development of engineering materials due to [1, 13, 29]

- high electrochemical discharge capacity,
- good electrochemical retaining rate during electrode life,
- fast hydrogen sorption kinetics,
- high hydrogen sorption capacity.

Studied chemical substitutions can be applied to other RE–Mg–Ni-type materials.

References

1. Jain, I. P., Lal, C., and Jain, A. (2010). Hydrogen storage in Mg: A most promising material. *Int. J. Hydrogen Energy* **35**: 5133–5144.
2. Zhao, X. G., and Ma, L. Q. (2009). Recent progress in hydrogen storage alloys for nickel/metal hydride secondary batteries—review. *Int. J. Hydrogen Energy* **34**: 4788–4796.
3. Zaluski, L., Zaluska, A., Tessier, P., Strom-Olsen, J. O., and Schulz, R. (1995). Catalytic effect of Pd on hydrogen absorption in mechanically alloyed Mg_2Ni, $LaNi_5$ and FeTi. *J. Alloys Comp.* **217**: 295–300.
4. Jurczyk, M., and Nowak, M. (2008). Nanostructured materials in electrochemistry, in: Eftekhari, A., eds., *Nanostructured Hydrogen Storage Materials Synthesized by Mechanical Alloying*, Weinheim: Wiley, pp. 349–385.
5. Jurczyk, M., Smardz, L., Okonska, I., Jankowska, E., Nowak, M., and Smardz, K. (2008). Nanoscale Mg-based materials for hydrogen storage. *Int. J. Hydrogen Energy* **33**: 374–380.
6. Varin, R. A., Czujko, T., Wronski, Z. S. (2009). *Nanomaterials for Solid State Hydrogen Storage*, Springer Science+Business Media, LLC.
7. Kadir, K., Sakai, T., and Uehara, I. (1997). Synthesis and structure determination of a new series of hydrogen storage alloys; RMg_2Ni_9 (R = La, Ce, Pr, Nd, Sm and Gd) built from $MgNi_2$ Laves-type layers alternating with AB_5 layers. *J. Alloys Comp.* **257**: 115–121.

8. Kohno, T., Yoshida, H., Kawashima, F., Inaba, T., Sakai, I., Yamamoto, M., et al. (2000). Hydrogen storage properties of new ternary system alloys: La$_2$MgNi$_9$, La$_5$Mg$_2$Ni$_{23}$, La$_3$MgNi$_{14}$. *J. Alloys Comp.* **311**: L5–L7.
9. Kohno, T., Yoshida, H., Kawashima, F., Inabat, T., Sakai, I., Yamamoto, M., and Kanda, M. (2000). Hydrogen storage properties of new ternary system alloys: La$_2$MgNi$_9$, La$_5$Mg$_2$Ni$_{23}$, La$_3$MgNi$_{14}$, *J. Alloys Comp.* **311** (2000) L5–L7.
10. Hayakawa, H., Akiba, E., Gotoh, M., and Kohno, T. (2005). Crystal structures of La-Mg-Ni$_x$ (x = 3–4) system hydrogen storage alloys. *Mater. Trans.* **46**: 1393–1401.
11. Zhang, F. L., Luo, Y. C., Chen, J. P., Yan, R. X., and Chen, J. H. (2007). La-Mg-Ni ternary hydrogen storage alloys with Ce$_2$Ni$_7$-type and Gd$_2$Co$_7$-type structure as negative electrodes for Ni/MH batteries. *J. Alloys Comp.* **430**: 302–307.
12. Chai, Y. J., Sakaki, K., Asano, K., Enoki, H., Akiba, E., and Kohno, T. (2007). Crystal structure and hydrogen storage properties of La-Mg-Ni-Co alloy with superstructure. *Scr. Mater.* **57**: 545–548.
13. Zhang, F. L., Luo, Y. C., Chen, J. P., Yan, R. X., Kang, L., and Chen, J. H. (2005). Effect of annealing treatment on structure and electrochemical properties of La$_{0.67}$Mg$_{0.33}$Ni$_{2.5}$Co$_{0.5}$ alloy electrodes. *J. Power Sources* **150**: 247–254.
14. Yartys, V. A., Riabov, A. B., Denys, R. V., Sato, M., and Delaplane, R. G. (2006). Novel intermetallic hydrides. *J. Alloys Comp.* **408**: 273–279.
15. Zhang, F. L., Luo, Y. C., Wang, D. H., Yan, R. X., Kang, L., and Chen, J. H. (2007). Structure and electrochemical properties of La$_{2-x}$Mg$_x$Ni$_{7.0}$ (x = 0.3–0.6) hydrogen storage alloys. *J. Alloys Comp.* **439**: 181–188.
16. Liu, Y., Cao, Y., Huang, L., Gao, M., and Pan, H. (2011). Rare earth-Mg-Ni-based hydrogen storage alloys as negative electrode materials for Ni/MH batteries. *J. Alloy. Comp.* **509**: 675–686.
17. Zhang, F. L., Luo, Y. C., Chen, J. P., Yan, R. X., and Chen, J. H. (2007). La-Mg-Ni ternary hydrogen storage alloys with Ce$_2$Ni$_7$-type and Gd$_2$Co$_7$-type structure as negative electrodes for Ni/MH batteries. *J. Alloys Comp.* **430**: 302–307.
18. Luo, Y., Jia, Q., Wu, T., and Kang, L. (2009). Influence of rare-earth elements R on phase-structure and electrochemical properties of hydrogen storage alloys (LaRMg) (NiCoAlZn)$_{3.5}$. *J. Lanzhou Univ. Technol.* **35**: 1–6.

19. Nakamura, J., Iwase, K., Hayakawa, H., Nakamura, Y., and Akiba, E. (2009). Structural study of La$_4$MgNi$_{19}$ hydride by in situ X-ray and neutron powder diffraction. *J. Phys. Chem. C* **113**: 5853–5859.

20. Zhang, Q. A., Fang, M. H., Si, T. Z., Fang, F., Sun, D. L., Ouyang, L. Z., et al. (2010). Phase stability, structural transition, and hydrogen absorption-desorption features of the polymorphic La$_4$MgNi$_{19}$ compound. *J. Phys. Chem. C* **114**: 11686–11692.

21. Liu, Y. F., Cao, H., Huang, L., Gao, M. X., and Pan, H. G. (2012). Rare earth–Mg–Ni-based hydrogen storage alloys as negative electrode materials for Ni/MH batteries. *J. Alloys Comp.* **509**: 675–686.

22. Chai, Y. J., Asano, K., Sakaki, K., Enoki, H., and Akiba, E. (2009). Phase transformation of the La$_{0.7}$Mg$_{0.3}$Ni$_{2.8}$Co$_{0.5}$-H$_2$ system studied by in situ X-ray diffraction. *J. Alloys Comp.* **485**: 174–80.

23. Denys, R. V., Yartys, V. A., Sato, M., Riabov, A. B., and Delaplane, R. G. (2007). Crystal chemistry and thermodynamic properties of anisotropic Ce$_2$Ni$_7$H$_{4.7}$ hydride. *J. Solid State Chem.* **180**: 2566–2576.

24. Ozaki, T., Kanemoto, M., Kakeya, T., Kitano, Y., Kuzuhara, M., Watada, M., et al. (2007). Stacking structures and electrode performances of rare earth-Mg-Ni-based alloys for advanced nickel-metal hydride battery. *J. Alloys Comp.* **446**: 620–624.

25. Nakamura, J., Iwase, K., Hayakawa, H., Nakamura, Y., and Akiba, E. (2009). Structural study of La$_4$MgNi$_{19}$ hydride by in situ X-ray and neutron powder diffraction. *J. Phys. Chem. C* **113**: 5853–5859.

26. Nakamura, Y., Nakamura, J., Iwase, K., and Akiba, E. (2009). Distribution of hydrogen in metal hydrides studied by in situ powder neutron diffraction. *Nucl. Instrum. Meth. A* **600**: 297–300.

27. Guzik, M. N., Hauback, B. C., and Yvon, K. (2012). Hydrogen atom distribution and hydrogen induced site depopulation for the La$_{2-x}$Mg$_x$Ni$_7$–H system. *J. Solid State Chem.* **186**: 9–16.

28. Nowak, M. (2016). Unpublished results.

29. Zhang, F. L., Luo, Y. C., Chen, J. P., Yan, R. X., and Chen, J. H. (2007). La-Mg-Ni ternary hydrogen storage alloys with Ce$_2$Ni$_7$-type and Gd$_2$Co$_7$-type structure as negative electrodes for Ni/MH batteries, *J. Alloys Comp.* **430**: 302–307.

30. Liu, Y. F., Cao, Y. H., Huang, L., Gao, M. G., and Pan, H. G. (2011). Rare earth–Mg–Ni-based hydrogen storage alloys as negative electrode materials for Ni/MH batteries-review. *J. Alloys Comp.* **509**: 675–686.

31. Jurczyk, M., Nowak, M., Smardz, L., and Szajek, A. (2011). Mg-based nanocomposites for room temperature hydrogen storage. *TMS Ann. Meeting* **1**: 229–236.

32. Balcerzak, M., Jakubowicz, J., Kachlicki, T., Jurczyk, M. (2015). Effect of multi-walled carbon nanotubes and palladium addition on the microstructural and electrochemical properties of the nanocrystalline Ti$_2$Ni alloy. *Int. J. Hydrogen Energy*, **40**: 3288–3299.

33. Balcerzak, M., Nowak, M. and Jurczyk, M. (2017). Hydrogenation and electrochemical studies of La-Mg-Ni alloys, *Int. J. Hydrogen Energy*, **42**: 1436–1443.

34. Liu, Y., Cao, Y., Huang, L., Gao, M., and Pan, H. (2011). Rare earth–Mg–Ni-based hydrogen storage alloys as negative electrode materials for Ni/MH batteries. *J. Alloys Comp.* **509**: 675–686.

35. Liao, B., Lei, Y. Q., Chen, L. X., Lu, G. L., Pan, H. G., and Wang, Q. D. (2004). Effect of the La/Mg ratio on the structure and electrochemical properties of La$_x$Mg$_{3-x}$Ni$_9$ (x = 1.6–2.2) hydrogen storage electrode alloys for nickel-metal hydride batteries. *J. Power Sources* **129**: 358–367.

36. Dong, X. P., Lu, F. X., Zhang, Y. H., Yang, L. Y., and Wang, X. L. (2008). Effect of La/Mg on the structure and electrochemical performance of La-Mg–Ni system hydrogen storage electrode alloy. *Mater. Chem. Phys.* **108**: 251–256.

37. Li, Y., Han, S. M., Li, J. H., Zhu, X. L., and Hu, L. (2008). The effect of Nd content on the electrochemical properties of low-Co La–Mg–Ni-based hydrogen storage alloys. *J. Alloys Comp.* **458**: 357–362.

38. Young, K., Ouchi, T., and Huang, B. (2014). Effects of various annealing conditions on (Nd, Mg, Zr)(Ni, Al, Co)$_{3.74}$ metal hydride alloys. *J. Power Sources* **248**: 147–153.

39. Liu, J. G., Li, Y., Han, D., Yang, S. Q., Chen, X. C., Zhang, Lu and Han, S. M. (2015). Electrochemical performance and capacity degradation mechanism of single-phase La-Mg-Ni based hydrogen storage alloys. *J. Power Sources* **300**: 77–86.

40. Si, T. Z., Zhang, Q. A., and Liu, N. (2008). Investigation on the structure and electrochemical properties of the laser sintered La$_{0.7}$Mg$_{0.3}$Ni$_{3.5}$ hydrogen storage alloys. *Int. J. Hydrogen Energy* **33**: 1729–1734.

41. Zhu, M., Peng, C. H., Ouyang, L. Z., and Tong, Y. Q. (2006). The effect of nanocrystalline formation on the hydrogen storage properties of AB$_3$-base Ml–Mg–Ni multi-phase alloys. *J. Alloys Comp.* **426**: 316–321.

42. Lide, D. R. (2009). *CRC Handbook of Chemistry and Physics*, 89th ed., CRC Press/Taylor and Francis, Boca Raton, FL.
43. Yasuoka, S., Magari, Y., Murata, T., Tanaka, T., Ishida, J., Nakamura, H., Nohma, T., Kihara, M., Baba, Y., and Teraoka, H. (2006). Development of high-capacity nickel-metal hydride batteries using superlattice hydrogen-absorbing alloys. *J. Power Sources* **156**: 662–666.

Chapter 15

Ni-MH$_x$ Batteries

Mieczyslaw Jurczyk and Marek Nowak

*Institute of Materials Science and Engineering,
Poznan University of Technology, Poznan, Poland*
mieczysław.jurczyk@put.poznan.pl, marek.nowak@put.poznan.pl

15.1 Introduction

A battery is a combination of two or more electrochemical cells which store chemical energy and make it available as electrical energy. Since its invention in 1800 by Alessandro Volta, the battery has become a common power source for many applications.

The voltage developed across a cell's terminals depends on the chemicals used in it and their respective concentrations. For example, alkaline and carbon-zinc cells both measure approximately 1.5 V, due to the energy release of the associated chemical reactions. Because of the high electrochemical potential changes in the reactions of lithium compounds, lithium cells can provide as much as 3 V or more (Table 15.1).

There are two main types of batteries, and each of them has its own advantages and disadvantages [1]:

- Primary batteries irreversibly (within limits of practicality) transform chemical energy to electrical energy. When the

---

Handbook of Nanomaterials for Hydrogen Storage
Edited by Mieczyslaw Jurczyk
Copyright © 2018 Pan Stanford Publishing Pte. Ltd.
ISBN 978-981-4745-66-6 (Hardcover), 978-1-315-36444-5 (eBook)
www.panstanford.com

initial supply of reactants is exhausted, energy cannot be readily restored to the battery by electrical means,
- Secondary batteries can be recharged, that is, they can have their chemical reactions reversed by supplying electrical energy to the cell, restoring their original composition.

Table 15.1 Rechargeable batteries selected for application in electric vehicles

Technology	Environmental	Nominal voltage [V]	Specific energy [Wh/kg]	Cycle life
Lead-acid	Toxic	2.0	20–35	100–500
Ni-Metal hydride	Low toxicity	1.2	55–70	500
Ni-cadmium	Toxic	1.2	30–50	1000
Sodium-sulfur ($T > 300°C$)	Hazardous	2.0	170	1000
Zinc-air	Nontoxic	1.5	70–85	600
Lithium-ion	Hazardous	1.5–3.9	115–150	200–500

Other portable rechargeable batteries include several "dry cell" types, which are sealed units and are therefore useful in appliances such as mobile phones and laptop computers. Cells of this type include nickel-cadmium (Ni-Cd), nickel-metal hydride (Ni-MH$_x$) and lithium-ion (Li-ion) cells. By far, Li-ion has the highest share of the dry cell rechargeable market. Meanwhile, Ni-MH$_x$ has replaced Ni-Cd in most applications due to its higher capacity, but Ni-Cd remains in use in power tools or medical equipment [2].

15.2 The Fundamental Concept of Hydride Electrode and Ni-MH$_x$ Battery

The interaction of hydrogen and metal M is represented as [3]

$$M + X/2H_2 \leftrightarrow MH_x(s), \qquad (15.1)$$

where MH$_x$ is the hydride of metal M.

Some alloys can be charged and discharged electrochemically. Equation 15.2 shows the electrochemical charging and discharging reactions:

$$M + xH_2O + xe^- \leftrightarrow MH_y + OH^- \tag{15.2}$$

The electrochemical reaction of a Ni-MH$_x$ cell can be represented by the following half-cell reactions. During charging, the nickel hydroxide Ni(OH)$_2$ positive electrode is oxidized to nickel oxyhydroxide NiOOH, while the alloy M negative electrode forms MH by water electrolysis. The reactions on each electrode proceed via solid-state transitions of hydrogen. The overall reaction is expressed only by a transfer of hydrogen between alloy M and Ni(OH)$_2$:

Nickel positive electrode

$$Ni(OH)_2 + OH^- = NiOOH + H_2O + e^- \tag{15.3}$$

Hydride negative electrode

$$M + H_2O + e^- = MH + OH^- \tag{15.4}$$

Overall reaction

$$Ni(OH)_2 + M = NiOOH + MH \tag{15.5}$$

In sealed Ni-MH$_x$ cell, the M electrode has a higher capacity than the Ni(OH)$_2$ electrode, thus facilitating a gas recombination reaction. During an overcharge situation, the MH electrode is charged continuously, forming hydride, while the Ni electrode starts to evolve oxygen gas according to Eq. (15.6):

Nickel positive electrode

$$2OH^- = H_2O + \frac{1}{2}O_2 + 2e^- \tag{15.6}$$

Hydride negative electrode

$$2M + 2H_2O + 2e^- = 2MH + 2OH^- \tag{15.7}$$

$$2MH + \frac{1}{2}O_2 = 2M + H_2O \tag{15.8}$$

Actual (non-ideal) case

$$H_2O + e^- = \frac{1}{2}H_2 + OH^- \quad (15.9)$$

$$2MH + \frac{1}{2}O_2 = 2M + H_2O \quad (15.10)$$

The oxygen diffuses through the separator to the MH electrode and there it reacts chemically, producing water (Eq. 15.10) and preventing a pressure rise in the cell.

During the over-discharge process, hydrogen gas starts to evolve at the Ni electrode (Eq. 15.11). The hydrogen diffuses through the separator to the MH electrode and there it dissociates to atomic hydrogen by a chemical reaction (Eq. 15.12), followed by a charge transfer reaction (Eq. 15.13), ideally causing no pressure rise in the cell:

Nickel positive electrode

$$2H_2O + 2e^- = H_2 + 2OH^- \quad (15.11)$$

Hydride negative electrode

$$H_2 + 2M = 2MH \quad (15.12)$$

$$2OH^- + 2MH = 2H_2O + 2e^- + 2M \quad (15.13)$$

15.3 Electrode Materials for Ni-MH$_x$ Batteries

In the Ni-MH$_x$ battery, the M/MH$_x$ electrode should be capable of reversible storage of hydrogen and should also exhibit an insignificant self-discharge. Since the reversibility of the charge-discharge reactions is required, a moderately stable hydride bonding between hydrogen and metal is desired. It is undesirable for the Me-H bond to be highly stable or very unstable. Because of this, detailed information about the energetics and kinetics of hydrogen transfer processes occurring on/in electrode materials, as well as the specific energy, charge-discharge efficiency and cyclic lifetime of metal-hydride electrodes, are of a great importance for their future use in commercial batteries (see Tables 15.2 and 15.3) [2, 4–27].

Table 15.2 The properties of AB_5-type electrodes in Ni-MH$_x$ battery in function of their chemical composition

Composition	Elements from periodic table and their influence on the properties
Substitutions of A in AB_5	$La_{1-x}M_xB_5$: Zr, Ce, Pr, and Nd improve activation, high-rate discharge and cycle life, but increase the self-discharge owing to a higher dissociation pressure of the metal hydride
Substitutions of B in AB_5	$A(Ni_{1-z}M_z)_5$: A = La, Mm; M = Co, Cu, Fe, Mn, Al; Co decreases the corrosion rate and improves the cycle life of the electrode, especially at elevated temperature (40°C), but increases the alloy costs. Substitution of Co by Fe allows cost reduction without affecting cell performance, decreases decrepitation of alloy during hydriding. Al increases hydride formation energy, prolongs cyclic life. Mn decreases equilibrium pressure without decreasing the amount of stored hydrogen. V increases the lattice volume and enhances the hydrogen diffusion. Cu increases high rate discharge performance
Additions to B in AB_5	$A(Ni, M)_{5-x}B_x$: A = La, Mm; M = Co, Cu, Fe, Mn, and Al; B = Al, Si, Sn, Ge, In, and Tl. The metals Al, Si, Sn, and Ge minimize corrosion of the hydride electrode. Ge-substituted alloys exhibit facilitated kinetics of hydrogen absorption/desorption in comparison with Sn-containing alloys. In, Tl, and Ga increase the overvoltage of hydrogen evolution (prevent generation of gaseous hydrogen)
Non-stoichiometric alloys	$AB_{5\pm x}$: A = La, Mm; B = (Ni, Mn, Al, Co, V, Cu). Additional Ni forms separate finely dispersed phase. In $MmB_{5.12}$, the Ni_3Al-type second phase with high electrocatalytic activity is formed. Alloys poor in Mm are destabilized and the attractive interaction between the dissolved hydrogen atoms increases. Second phase (Ce_2Ni_7), which forms very stable hydride, is present in $MmB_{4.88}$. When $(5-x) < 4.8$ the hydrogen gas evolution during overcharge decreases
Mixture of two AB_5-type alloys	Electrode performance can be improved by mixing of two alloys characterized by various hydrogen equilibrium absorption pressures.

Table 15.3 The properties of the AB_2- and AB-type electrodes in Ni-MH$_x$ battery in function of their chemical composition

Composition	Elements from periodic table and their influence on the properties
ZrM_2-type alloy (M = V, Cr, Mn, Ni)	$Zr(V_xNi_{1-x})_2$: An increase of the V content increases the maximum amount of absorbed hydrogen. Ni substitution decreases the electrochemical activity of an alloy. Composite alloy, mixture of $ZrNi_2$ with RNi_5 (R = rare earth element), shows improved characteristics in comparison with the parent compounds
Over stoichiometric ZrB_2-type alloy	$ZrNi_{1.2}Mn_{0.6}V_{0.2}Cr_{0.1}X_x$ (X = La or Ce; x = 0.05): La and Ce improve the activation behavior of alloy during chemical pretreatment and increase discharge capacity $ZrV_{1.5}Ni_{1.5}$: The over-stoichiometric alloy, in which some of the V atoms move from B to A sites, shows very high capacity
TiFe-type alloy	$TiFe_{1-x}Pd_x$: Pd substitution increases both the lattice constant and the catalytic activity, decreases the plateau pressure
Vanadium-based alloys	$V_3Ti(Ni_{1-x}M_x)$, $V_3TiNi_xM_y$: M = Al, Si, Mn, Fe, Co, Cu, Ge, Zr, Nb, Mo, Pd, Hf, and Ta. The addition of Co, Nb, and Ta improves cycling durability. Alloys with Hf (or Nb), Ta and Pd show higher discharge capacity. M = Hf improves the high rate capability. TiNi phase exhibits high electrocatalytic activity
Substitution of A in AB_2	A = Zr + Ti: $Ti_xZr_{1-x}Ni_{1.1}$, $V_{0.5}Mn_{0.2}Fe_{0.2}$: Electrodes without Ti, or a very low Ti content, exhibit excellent cycling and electrochemical stability. Electrodes with Ti:Zr atomic ratio 2:1 display higher storage capacity than that with Ti:Zr = 1:1, and higher electrochemical activity. Amount of C15 phase decreases with increasing x; at x = 0.5 it is pure C14 phase, at x = 0.75 it is 13% cubic bcc phase + 87% C14 phase; bcc phase absorbs more hydrogen than the C14 hexagonal phase

Composition	Elements from periodic table and their influence on the properties
Substitution of B in AB_2	$Zr_{0.8}Ti_{0.2}(V_{0.3}Ni_{0.6}M_{0.1})_2$, $Ti_{0.35}Zr_{0.65}Ni_{1.2}V_{0.6}Mn_{0.2}Cr_{0.2}$: The Si, Mn-substituted alloys has C14 Laves phase structure. The Co, Mo-substituted alloys form Cl5 Laves phase structure. Mn enhances the activation of an alloy during chemical pretreatment and increases discharge capacity. Co addition leads to the longest cyclic lifetime. Cr addition reduces the discharge capacity but extends cyclic lifetime; Cr controls the dissolution of V and Zr
Non-stoichiometric alloys $AB_{2\pm x}$	$Zr_{0.495}Ti_{0.505}V_{0.771}Ni_{1.546}$, $Ti_{0.8}Zr_{0.2}Ni_{0.6}V_{0.64}Mn_{0.4}$, $ZrV_{0.8}Ni_{45}M_5$: Increasing the Ni content C14-type Laves phase preserves an over-stoichiometric alloy and discharge capacity increases on increasing the amount of Ni decreases V-rich dendrite formations
MgNi-based alloys for electrodes	Ternary alloy: $Mg_{50}Ni_{45}M_5$ (M = Mn, Cu, Fe): Ni substitution with Zn increases the deterioration rate. Substitution of Fe, W, Cu, Mn, Cr, Al, or C instead of Ni decreases both the deterioration rate and discharge capacity. Ni substitution by Se, Cu, Co, or Si decreases both the discharge capacity and cycling life

As it was mentioned earlier conventionally, the microcrystalline hydride materials have been prepared by arc or induction melting and annealing (see Section 4.2). The arc melting technique provides two main advantages over other melt processes: great versatility in terms of the types of materials, and limited reactivity during melting. Disadvantages of this process are low efficiency and hazardous preparation conditions as well as in the case of multicomponent alloys, multiple remelting or prolonged annealing at high temperatures. On the other hand, the induction melting process is used for large-scale alloy production.

However, either a low storage capacity by weight or poor absorption-desorption kinetics in addition to a complicated activation procedure have limited the practical use of microcrystalline metal hydrides. Substantial improvements in the hydriding-dehydriding properties of metal hydrides could be possibly achieved by the formation of nanocrystalline structures by non-equilibrium processing techniques such as mechanical alloying (MA) or high-energy ball-milling (HEBM) [28–34].

Until now, various performance parameters of the Ni-MH$_x$ batteries such as electrochemical hydrogen storage capacity and cycle life were investigated and compared with theoretical hydrogen storage capacity (see Table 15.4). It is important to note that the electrode made from LaNi$_5$-phase reaches maximum capacity (360 mAh/g) in the first cycle, but the discharge capacity decreases quickly in the next cycles [35]. The properties of hydrogen host material can be modified substantially by alloying, to obtain the desired storage characteristics, e.g., proper capacity at a favorable hydrogen pressure. For example, it was found that the partial respective replacement of Ni in LaNi$_5$ by small amounts of Al resulted in a prominent increase in the cycle lifetime without causing much decrease in capacity [36]. Aluminum is believed to concentrate on grain boundaries and in connection with segregated La forms a porous oxide layer, which protects the material from further corrosion in KOH electrolyte. On the other hand, cobalt is contained in the alloys to guarantee the long cycle life of the negative electrode [37]. These electrodes usually obtain their maximum capacity within a few charge-discharge cycles without any special pretreatment. The substitution of Mn provided the LaNi$_5$ alloy with a more brittle character, giving a shorter cycle life [38]. Generally, in the transition metal sublattice of LaNi$_5$-type compounds substitution by Mn, Al and Co has been found to offer the best compromise between high hydrogen capacity and good resistance to corrosion [39].

In 1998, Iwakura et al. investigated the effect of partial displacement of cobalt for nickel by measuring the crystallographic, thermomagnetic and kinetic properties of hydrogen storage alloys in the composition range of microcrystalline MmNi$_{4.0-x}$Mn$_{0.75}$Al$_{0.25}$Co$_x$ ($0 \leq x \leq 0.6$) [40]. Increase in cobalt content of the alloys, due to the enlargement of the unit cell volume, stabilizes the resulting hydride and improves the cycling durability without a decrease in discharge capacity.

In the case of AB$_2$-type electrode alloys, the electrochemical activity of these materials can be stimulated by partial substitution in which Zr is partially replaced by Ti, and V is partially replaced by other transition metals (Cr, Mn and Ni) [28, 41–43]. Some microcrystalline alloys with non-stoichiometric compositions, such as ZrV_{2+x}-type, have also been studied [44].

Table 15.4 Electrochemical characteristics of selected hydrogen-absorbing alloys

Composition of alloy	Theoretical hydrogen storage capacity [mAh/g]	Low rate discharge capacity [mAh/g]	High rate to low rate discharge capacity [%]
AB_5-type alloys			
$MmNi_4Al$	270	240	82
$La_{0.8}Ce_{0.2}Ni_2Co_3$	260	253	39
$MmNi_{3.5}Co_{0.7}Al_{0.8}$	240	167	50
$LaNi_{5.4}$	360	170	25
$LaNi_4Cu$	373	270	11
$LaNi_{4.2}Cu$	360	260	62
$LaNi_5Cu$	198	180	98
$LaNi_5$	370	171	95
AB_2-type alloys			
$ZrV_{0.2}Mn_{0.8}Ni$-C14		190	20
$ZrV_{0.3}Mn_{0.7}Ni$-C14, C15		240	34
$ZrV_{0.4}Mn_{0.6}Ni$-C14, C15		290	60
$ZrV_{0.5}Mn_{0.5}Ni$-C14, C15		280	82
$Ti_{0.26}Zr_{0.07}V_{0.24}Mn_{0.20}Ni_{0.23}$		295	—
AB-type alloys			
$Ti_{0.35}Zr_{0.65}NiV_{0.6}Mn_{0.4}$		305	25
$Ti_{0.35}Zr_{0.65}Ni_{1.2}V_{0.6}Mn_{0.2}$		305	52
$Ti_{0.35}Zr_{0.65}Ni_{1.2}V_{0.4}Mn_{0.4}$		260	54
Mg-Ni alloys			
Mg_2Ni	999	320 (30°C) 421 (90°C)	
$Mg_{1.9}Al_{0.1}Ni_{0.9}Y_{0.1}$		150	70

Among the different types of hydrogen forming compounds, Ti-based alloys are among the promising materials for hydrogen energy applications [30, 45]. Nevertheless, the application of TiFe material in batteries has been limited due to poor absorption/ desorption kinetics in addition to a complicated activation procedure. To improve the activation of this alloy several approaches have been adopted. For example, the replacement of Fe by some amount of transition metals to form secondary phase may improve the activation property of TiFe [46, 47]. After activation, the TiFe reacts directly and reversibly with hydrogen to form two ternary hydrides TiFeH (orthorhombic) and TiFeH$_2$ (monoclinic), each of which is a distorted form of the bcc structure of the unhydrided alloy [48]. In addition, excess Ti in TiFe, i.e., Ti$_{1+x}$Fe enables the alloy to be hydrided without the activation treatment. On the other hand, ball-milling of TiFe is effective for the improvement of the initial hydrogen absorption rate, due to the reduction in the particle size and to the creation of new clean surfaces [49, 50].

It has been shown that Ti-based alloys are among the most promising materials for the electrode materials in nickel-metal hydride batteries [51, 52]. Titanium and iron form two stable intermetallic compounds, TiFe and TiFe$_2$. TiFe$_2$ does not absorb hydrogen. The TiFe alloy, which crystallizes in the cubic CsCl-type structure, is cheaper and lighter than the LaNi$_5$-type alloys and can absorb up to 2 H/f.u. at room temperature. This alloy forms two hydrides, TiFeH (orthorhombic phase) and TiFeH$_2$ (monoclinic phase). The discharge capacity of conventionally produced (e.g., arc melted, annealed and mechanically crushed) TiFe powder is 0 mA h g^{-1} [53]. To improve the activation of this microcrystalline alloy several approaches have been adopted. Recently, it has been shown that the discharge capacity of TiFe increased from 0 to 64 mA h g^{-1} after high-energy ball-milling [53]. A fine nanostructure is known to improve greatly the kinetics and activation of hydrogen storage alloys, when compared with their conventional polycrystalline counterparts. On the other hand, the substitution of Fe by some amount of transition metals may improve the activation property of TiFe. For example, with increasing nickel content in nanocrystalline TiFe$_{1-x}$Ni$_x$ the material shows an increase in discharge capacity which passes through a maximum for $x = 0.75$. In the nanocrystalline TiFe$_{0.25}$Ni$_{0.75}$ powder, discharge capacities of up to 155 mA h g^{-1} (at 40 mA g^{-1} discharge current) were measured [52, 54, 55].

Magnesium-based alloys have been extensively studied during last years, but the microcrystalline Mg$_2$Ni alloy can reversibly absorb and desorb hydrogen only at high temperatures. Substantial improvements in the hydriding-dehydriding properties of Mg-type metal hydrides could possibly be achieved by formation of nanocrystalline structures [28, 29, 56, 57]. It was found that the electrochemical activity of nanocrystalline hydrogen storage alloys can be improved in many ways, by alloying with other elements, by ball-milling the alloy powders with a small amount of nickel or graphite powders [58–63]. For example, the surface modification of nanocrystalline hydrogen storage alloys with graphite by ball-milling leads to an improvement in both discharge capacity and charge-discharge cycle life [62, 63].

15.4 Sealed Ni-MH$_x$ Batteries

The Ni/MH battery was discovered in the 1970s and introduced into market in the 1990s [64]. It is a new type of green rechargeable battery with a nickel hydroxide electrode as its positive electrode, a hydrogen storage alloy electrode as its negative electrode and a potassium hydroxide (KOH) solution as its electrolyte. Figure 15.1 a shows schematically the electrochemical charge–discharge process of a Ni/MH battery. The overall electrochemical reaction of a Ni/MH battery can be described as follows:

$$x\text{Ni(OH)}_2 + \text{M} \underset{\text{discharge}}{\overset{\text{charge}}{\longleftrightarrow}} x\text{NiOOH} + \text{MH}_x$$

Upon charging, hydrogen atoms dissociate from Ni(OH)$_2$ at the positive electrode and are absorbed by the hydrogen storage alloy to form a metal hydride at the negative electrode. Upon discharging, the hydrogen atoms stored in the metal hydride dissociate at the negative electrode and react with NiOOH to form Ni(OH)$_2$ at the positive electrode. Therefore, the charge–discharge mechanism for a Ni-MH$_x$ battery is merely a movement of hydrogen between a metal hydride electrode and a nickel hydroxide electrode in an alkaline electrolyte, viz., the "rocking-chair" mechanism.

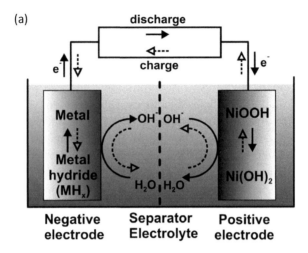

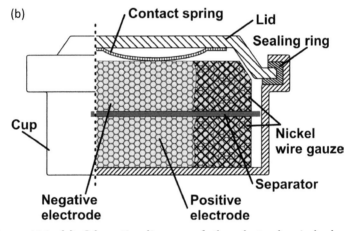

Figure 15.1 (a) Schematic diagram of the electrochemical charge–discharge reaction process of a Ni/MH battery, (b) sealed cell.

The cyclic behavior of the some nanostructured alloy anodes was examined in a sealed HB 116/054 cell (according to the International standard IEC no. 61808, related to the hydride button rechargeable single cell) (Fig. 15.1) [65, 66]. The mass of the active material was 0.33 g. To prepare MH negative electrodes, alloy powders were mixed with addition of 5 wt% tetracarbonyl nickel. Then this mixture was pressed into the form tablets, which were placed in a small basket made of nickel nets (as the current collector). The diameter of each tested button cells was

6.6 mm and a thickness of 2.25 mm, respectively. The sealed Ni-MH cell was constructed by the pressing negative and positive electrode, polyamide separator and KOH (ρ = 1.20 × 10^{-3} kg m^{-3}) as electrolyte solution. The battery with electrode fabricated from nanocrystalline materials was charged at current density of i = 3 mA g^{-1} for 15 h and after 1 h pause discharged at current density of i = 7 mA g^{-1} down to 1.0 V. All electrochemical measurements were performed at 20 ± 1°C.

To study the quality of the activated $TiFe_{0.25}Ni_{0.75}$ and TiNi as electrode materials in the Ni-MH_x battery, the overpotential dependence on the current density (i = 10, 20, 40, and 80 mA g^{-1}) was recorded at 15 s of anodic and cathodic galvanostatic pulses (Fig. 15.2). It can be see that the anodic and cathodic parts of nanocrystalline electrodes are almost symmetrical with respect to the rest potential of the electrode. It may be concluded that for all studied electrodes, fast rates can be achieved.

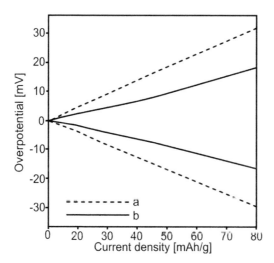

Figure 15.2 Overpotential against current density on activated nanocrystalline: (a) $TiFe_{0.25}Ni_{0.75}$; and (b) TiNi electrodes, at 15 s of anodic and cathodic galvanostatic pulses.

Figure 15.3 shows the discharge capacities of sealed button cells with electrodes prepared from nanocrystalline Ti-based alloys as a function of discharge cycle number. Among TiNi-type materials, the highest discharge capacities were found for $TiFe_{0.25}Ni_{0.75}$ alloy. It is interesting to note that that the sealed battery using the nanocrystalline TiFe$_{0.25}$Ni$_{0.75}$ alloy has almost the capacity of the microcrystalline (Zr$_{0.35}$Ti$_{0.65}$) (V$_{0.93}$Cr$_{0.28}$Fe$_{0.19}$Ni$_{1.0}$) one [67].

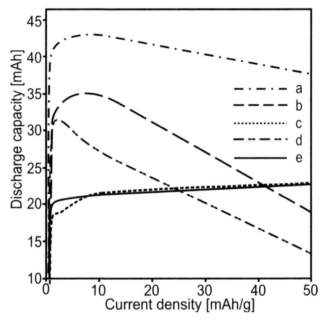

Figure 15.3 Durability of the sealed button cells with negative electrodes made from nanocrystalline Ti-based alloys: (a) TiFe$_{0.25}$Ni$_{0.75}$, (b) TiNi, (c) TiNi$_{0.875}$Zr$_{0.125}$, (d) TiNi$_{0.6}$Fe$_{0.1}$Mo$_{0.1}$Cr$_{0.1}$Co$_{0.1}$, and (e) NiCd alloys (the mass of the active material was 0.33 g).

Independently, it was found that the discharge capacities of sealed button batteries with electrodes prepared from the nanocrystalline La(Ni, Mn, Al, Co)$_5$ powders had a slightly higher discharge capacities, compared with negative electrodes prepared from microcrystalline powders [68]. Figures 15.4 and 15.5 shows the discharge capacities of sealed button cells with electrodes prepared from MmNi$_{3.5}$Al$_{0.8}$Co$_{0.7}$ as well as Zr$_{0.975}$La$_{0.025}$V$_{0.56}$Ni$_{1.69}$ alloy powders as a function of discharge cycle, respectively. It is interesting to note that the amorphous MmNi$_{3.5}$Al$_{0.8}$Co$_{0.7}$ and Zr$_{0.975}$La$_{0.025}$V$_{0.56}$Ni$_{1.69}$ alloys are not interesting to sealed battery application due to very low discharge capacities.

Sealed Ni-MH$_x$ Batteries | 293

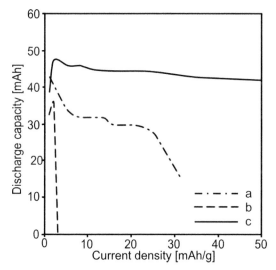

Figure 15.4 Discharge capacity as a function of cycle number of electrode prepared with polycrystalline (a), amorphous (b) and nanocrystalline MmNi$_{3.5}$Al$_{0.8}$Co$_{0.7}$ (c); (solution, 6M KOH; temperature, $T = 23°C$). The charging current 4 mAg^{-1}; the discharge current 8 mAg^{-1}; The final discharge voltage of 1 V.

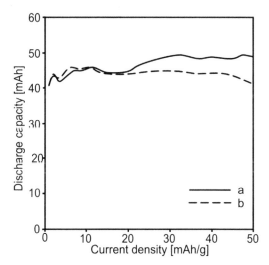

Figure 15.5 Discharge capacity as a function of cycle number of electrode prepared with polycrystalline (a) and nanocrystalline Zr$_{0.975}$La$_{0.025}$V$_{0.56}$Ni$_{1.69}$ (b); (solution, 6M KOH; temperature, $T = 23°C$). The charging current 4 mAg^{-1}; the discharge current 8 mAg^{-1}; The final discharge voltage of 1 V.

15.5 A Composite Hydrogen Storage Alloy in Application in Sealed Ni-MH$_x$ Batteries

Recently, composite hydrogen storage alloys based on Mg or containing Mg due to high hydrogen storage capacity, low weight, low cost may be the most effective materials for practical application in new Ni-MH$_x$ batteries. This subchapter reviews the development of composite hydrogen storage alloys prepared by mechanical alloying and high-energy ball-milling.

Mechanical alloying has several important advantages for preparing hydrogen storage alloys [69, 70]:

- It allows easy and controlled synthesis of equilibrium and metastable alloy phases such as crystalline, nanocrystalline, amorphous, and quasicrystalline in spite of large differences in the melting points of the raw materials.
- It can effectively synthesize a composite hydrogen storage material which contains two or more hydrogen storage alloys.
- Its product is a powder which can be used directly for hydrogen storage materials without subsequent size reduction.
- It is a simple and inexpensive processing technology operating at room temperature.

A composite hydrogen storage alloy contains two or more hydrogen storage alloys, or a hydrogen storage alloy and another intermetallic compound. Generally, the major component in a composite hydrogen storage alloy is an alloy with good hydrogen storage properties. The minor component is a surface activator to improve the activation properties and the kinetics of hydrogen sorption/desorption [71]. Many approaches to surface modification have been attempted to dissolve the oxide layer and promote the formation of an active surface with excellent electrocatalytic performance [71–74].

Desirable electrochemical properties of hydrogen storage alloys include high storage capacity, easy activation, high resistance to corrosion, favorable kinetic performance, high rate dischargeability (HRD), and low cost. In practice, it is difficult to simultaneously obtain all these properties in a single alloy system, but using a composite hydrogen storage alloy is an effective way to achieve it. Table 15.5 summarizes the exchange current density (I_0) and hydrogen diffusion coefficient (D) values of some hydrogen storage alloys in a half-cell containing a Ni(OH)$_2$/NiOOH counter electrode and a Hg/HgO reference electrode in the KOH solution [69, 75].

Table 15.5 Exchange current densities for some hydrogen storage alloys at 50% depth of discharge (DOD)

Composition	Phase structure	Exchange current density, I_0 (mA/g)	Hydrogen diffusion coefficient, D (cm^2/s)
Ti$_{0.8}$Zr$_{0.2}$V$_{2.7}$Mn$_{0.5}$Cr$_{0.8}$Ni$_{1.25}$	C14 Laves phase, V-based solid solution phase	142.8	4.6×10^{-11}
Ti$_{45}$Zr$_{35}$Ni$_{17}$Cu$_3$	I-phase	199.5	5.8×10^{-10}
La$_2$MgNi$_9$	PuNi$_3$-type	88.5	—
La$_{0.7}$Mg$_{0.3}$Ni$_{3.5}$	PuNi$_3$-type, LaNi$_5$	79.6	7.7×10^{-10}
La$_{1.7}$Mg$_{0.3}$Ni$_{7.0}$	LaNi$_5$, PuNi$_3$-type	128.1	4.5×10^{-10}
Mg$_{0.86}$Ti$_{0.1}$Pd$_{0.04}$Ni	Amorphous	256	3.0×10^{-9}
La$_{0.7}$Mg$_{0.3}$Ni$_{3.5}$ + 5 wt% Ti$_{0.17}$Zr$_{0.08}$V$_{0.35}$Cr$_{0.1}$Ni$_{0.3}$	PuNi$_3$-type, LaNi$_5$, V-based solid solution phase, C14 Laves phase	362.9	1.67×10^{-11}

Charge-transfer and hydrogen diffusion expressed by the exchange current density (I_0) and hydrogen diffusion coefficient (D) are the two most important factors for evaluating the electrochemical properties of hydrogen storage alloys [69]. Currently, using composite hydrogen storage alloys may be most effective for practical battery applications.

15.6 Major Markets for Ni-MH$_x$ Batteries

Currently, there are three major markets for Ni-MH$_x$ batteries [76]: consumer electronics, hybrid electrical vehicle (HEV), and stationary energy storage units. In the first case, the rechargeable Ni-MH$_x$ battery was applied in portable electronic devices such as cell phones, cordless phones, smart vacuum cleaners, power tools, shavers, and toothbrushes. For HEV propulsion, the current battery market is dominated by the Ni-MH$_x$ battery. The market share of HEV battery sales by Toyota was over USD 2 billion in 2012 and increases yearly about 10%. In the stationary Ni-MH$_x$ energy market, the following applications can be found: uninterrupted power supply (UPS), energy storage for alternative energy generation (solar and winds farms), and buffer between alternative energy and the electrical grid.

References

1. *Handbook of Batteries* (2002). Linden, D., and Reddy, T. B. (Eds). 3rd edition. McGraw-Hill.
2. Kleparis, J., Wojcik, G., Czerwinski, A., Skowronski, J., Kopczyk, M., and Beltowska-Brzezinska, M. (2001). Electrochemical behavior of metal hydrides. *J. Solid State Electrochem.* **5**:229–249.
3. Hong, K. (2001). The development of hydrogen storage alloys and the progress of nickel hydride batteries. *J. Alloys Comp.* **321**:307–313.
4. Willems, J. J. G. (1984). Metal hydride electrodes stability of LaNi$_5$ related alloys. *Philips J. Res.* **39**:1–94.
5. Van Vucht, J. H. N., Kuijpers, F. A., and Bruning, H. C. A. M. (1970). Reversible room-temperature absorption of large quantities of hydrogen by intermetallic compounds. *Philips Res. Rep.* **25**:133–140.
6. Ivey, D. G., and Northwood, D. O. (1983). Storing energy in metal hydrides: a review of the physical metallurgy. *J. Mater. Sci.* **18**:321–347.
7. Furukawa, N. (1994). Development and commercialization of nickel metal hydride secondary batteries. *J. Power Sources* **51**:45–59.
8. Martin, M., Gommel, C., Borkhart, C., and Fromm, E. (1996). Absorption and desorption kinetics of hydrogen storage alloys. *J. Alloys Comp.* **238**:193–201.
9. Zheng, G., Popov, B. N., and White, R. E. (1996). Determination of transport and electrochemical kinetic parameters of bare and copper-coated LaNi$_{4.27}$Sn$_{0.24}$ electrodes in alkaline solution. *J. Electrochem. Soc.* **143**:835–839.
10. Zheng, G., Popov, B. N., and White, R. E. (1996). Application of porous electrode theory on metal hydride electrode in alkaline solution. *J. Electrochem. Soc.* **143**:435–441.

11. Lee, H. H., Lee, K. Y., and Lee, J. Y. (1996). The Ti-based metal hydride electrode for Ni-MH rechargeable batteries. *J. Alloys Comp.* **239**:63–70.
12. Wang, C. S., Lei, Y. Q., and Wang, Q. D. (1998). Studies of electrochemical properties of TiNi alloy used as an MH electrode. I. Discharge capacity. *Electrochim. Acta* **43**:3193–3207.
13. Züttel, A. (2003). Materials for hydrogen storage. *Mater. Today* **6**: 24–33.
14. Fernández, G. E., Rodríguez, D., and Meyer, G. (1998). Hydrogen absorption kinetics of $MmNi_{4.7}Al_{0.3}$. *Int. J. Hydrogen Energy* **23**: 1193–1196.
15. Srivastava, S., and Srivastava, O. N. (1999). Synthesis, characterization and hydrogenation behaviour of composite hydrogen storage alloys, $LaNi_5/La_2Ni_7$, $LaNi_3$. *J. Alloys Comp.* **282**:197–205.
16. Akiba, E. (1999). Hydrogen-absorbing alloys. *Curr. Opin. Solid St. M* **4**: 267–272.
17. Sakai, T., Uehara, I., and Ishikawa, H. (1999). R&D on metal hydride materials and Ni–MH batteries in Japan. *J. Alloys Comp.* **293–295**: 762–769.
18. Besenhard, J. O. (1999). *Handbook of Battery Materials*, 1st ed., Wiley–VCH, New York.
19. Boter, P. A. (1977). Rechargeable electrochemical cell, US Patent 4,004,943.
20. Hong, K. (2001). The development of hydrogen storage alloys and the progress of nickel hydride batteries. *J. Alloys Comp.* **321**:307–313.
21. Jankowska, E., and Jurczyk, M. (2002). Electrochemical behaviour of high-energy ball-milled TiFe alloy. *J. Alloys Comp.* **346**(1–2):L1–3.
22. Jurczyk, M., Jankowska, E., Nowak, M., and Jakubowicz, J. (2002). Nanocrystalline titanium-type metal hydride electrodes prepared by mechanical alloying. *J. Alloys Comp.* **336**(1–2):265–269.
23. Xu, Y. H., He, G. R., and Wang, X. L. (2003). Hydrogen evolution reaction on the AB_5 metal hydride electrode. *Int. J. Hydrogen Energy* **28**(9):961–965.
24. Feng, F., and Northwood, D. O. (2004). Hydrogen diffusion in the anode of Ni/MH secondary batteries. *J. Power Sources* **136**(2):346–350.
25. Han, S. M., Zhao, M. S., Zhang, Z., Zheng, Y. Z., and Jing, T. F. (2005). Effect of AB_2 alloy addition on the phase structures and electrochemical characteristics of $LaNi_5$ hydride electrode. *J. Alloys Comp.* **392**(1–2):268–273.

26. Drenchev, B., and Spassov, T. (2007). Electrochemical hydriding of amorphous and nanocrystalline TiNi-based alloys. *J. Alloys Comp.* **441**(1–2):197–201.
27. Zhao, X. G., and Ma, L. Q. (2009). Recent progress in hydrogen storage alloys for nickel/metal hydride secondary batteries—Review, *Inter. J. Hydrogen Energy* **34**:4788–4796.
28. Anani, A., Visintin, A., Petrov, K., and Srinivasan, S. (1994). Alloys for hydrogen storage in nickel/hydrogen and nickel/metal hydride batteries. *J. Power Sources* **47**:261–275.
29. Jurczyk, M. (2004). The progress of nanocrystalline hydride electrode materials. *Bull. Pol. Ac., Tech.* **52**:67–77.
30. Zaluski, L., Zaluska, A., Ström-Olsen, J. O. (1997). Nanocrystalline metal hydrides. *J. Alloys Comp.* **253–254**:70–79.
31. Zaluska, A., Zaluski, L., Ström-Olsen, J. O. (2001). Structure, catalysis, and atomic reactions on the nano-scale: A systematic approach to metal hydrides for hydrogen storage. *Appl. Phys. A* **72**:157–165.
32. Okonska, I., Jurczyk, M. (2009). Electrochemical properties of an amorphous 2Mg+3d alloys doped by nickel atoms (3d = Fe, Co, Ni, Cu). *J. Alloys Comp.* **475**:289–293.
33. Nowak, M., Okonska, I., Smardz, L., Jurczyk, M. (2009). Segregation effect on nanoscale Mg-based hydrogen storage materials. *Mater. Sci. Forum* **610–613**:472–479.
34. Jurczyk, M., Nowak, M., Okonska, I., Smardz, L., and Szajek, A. (2009). Nanocomposite hydride LaNi$_5$/A- and Mg$_2$Ni/A-type materials (A = C, Cu, Pd). *Mater. Sci. Forum* **610–613**: 472–479.
35. Zhao, X., and Ma, L. (2009). Recent progress in hydrogen storage alloys for nickel/metal hydride secondary batteries. *Int. J. Hydrogen Energy* **34**:4788–4796.
36. Nakamura, Y., Nakamura, H., Fujitani, S., Yonezu, I. (1994). Homogenizing behaviour in a hydrogen-absorbing LaNi$_{4.55}$Al$_{0.45}$ alloy through annealing and rapid quenching. *J. Alloys Comp.* **210**:299–303.
37. David, E. (2005). An overview of advanced materials for hydrogen storage. *J. Mater. Process. Technol.* **162–163**:169–177.
38. Sakai, T., Matsuoka, M., Iwakura, C. (1995). In *Handbook on the Physics and Chemistry of Rare Earths* (Gschneidner K. A., Eyring, L. eds), vol. 21. Elsevier, Amsterdam, pp. 135–180.
39. Joubert, J. M., Latroche, M., Percheron-Guégan, A., Bourée-Vigneron, F. (1998). Thermodynamic and structural comparison between two

potential metal-hydride battery materials $LaNi_{3.55}Mn_{0.4}Al_{0.3}Co_{0.75}$ and $CeNi_{3.55}Mn_{0.4}Al_{0.3}Co_{0.75}$. *J. Alloys Comp.* **275–277**:118–122.

40. Iwakura, C., Fukuda, K., Senoh, H., Inoue, H., Matsuoka, M., and Yamamoto, Y. (1998). Electrochemical characterization of $MmNi_{4.0-x}Mn_{0.75}Al_{0.25}Co_x$ electrodes as a function of cobalt content. *Electrochim. Acta* **43**:2041–2046.

41. Anani, A., Visintin, A., Srinivasan, S., Appleby, A. J., Reilly, J. J., Johnson, J. R. (1992). In *Proceedings of the Symposium on Hydrogen Storage Materials, Batteries, and Electrochemistry*; Corrigan, D. A., Srinivasan, S. Eds.; Pennington; NJ, Electrochemical Society, vol. 92-5, 1992, p. 105.

42. Nakano, H., and Wakao, S. (1995). Substitution effect of elements in Zr-based alloys with Laves phase for nickel-hydride battery. *J. Alloy. Compd.* **231**:587–593.

43. Kopczyk, M., Wojcik, G., Młynarek, G., Sierczynska, A., and Beltowska-Brzezinska, M. (1996). Electrochemical absorption-desorption of hydrogen on multicomponent Zr-Ti-V-Ni-Cr-Fe alloys in alkaline solution. *J. Appl. Electrochem.* **26**:639–645.

44. Tsukahara, M., Takahashi, K., Mishima, T., Isomura, A., and Sakai, T. (1996). V-based solid solution alloys with Laves phase network: Hydrogen absorption properties and microstructure. *J. Alloys Comp.* **236**:151–155.

45. Buschow, K. H. J., Bouten, P. C. P., and Miedema, A. R. (1982). Hydrides formed from intermetallic compounds of two transition metals—a special class of ternary alloys. *Rep. Prog. Phys.* **45**:937–1039.

46. Luan, B., Cui, N., Liu, H. K., Zhao, H. J., and Dou, S. X. (1995). Effect of cobalt addition on the performance of titanium-based hydrogen-storage electrodes. *J. Power Sources* **55**:197–203.

47. Lee, S. M., and Perng, T. P. (1999). Correlation of substitutional solid solution with hydrogenation properties of $TiFe_{1-x}M_x$(M=Ni, Co, Al) alloys. *J. Alloys Comp.* **291**:254–261.

48. Bradhurst, D. H. (1983). Metal hydrides for energy storage. *Metals Forum* **6**:139–148.

49. Schulz, R., Huot, J., Liang, G., Boily, S., Lalande, G., Denis, M, C., et al. (1999). Recent development in the applications of nanocrystalline materials to hydrogen technologies. *Mater. Sci. Eng. A* **267**:240–245.

50. Jung, C. B., and Lee, K. S. (1997). Electrode characteristics of metal hydride electrodes prepared by mechanical alloying. *J. Alloys Comp.* **253–254**:605–608.

51. Zaluski, L., Zaluska, A., Tessier, P., Strom-Olsen, J. O., and Schulz, R. (1995). Catalytic effect of Pd on hydrogen absorption in mechanically alloyed Mg$_2$Ni, LaNi$_5$ and FeTi, *J. Alloys Comp.* **217**:295–300.

52. Nowak, M., and Jurczyk, M. (2017). Nanotechnology for the Storage of Hydrogen. In *Nanotechnology for Energy Sustainability*, 3 Vol., Raj, B., Van de Voorde, M. and Mahajan, Y. (Eds) ISBN: 978-3-527-34014-9.

53. Jankowska, E., and Jurczyk, M. (2002). Electrochemical behaviour of high-energy ball-milled TiFe alloy. *J. Alloys Comp.* **346**:L1–L3.

54. Jurczyk, M., Nowak, M., Jankowska, E., and Jakubowicz, J. (2002). Structure and electrochemical properties of the mechanically alloyed La(Ni, M)$_5$ materials. *J. Alloys Comp.* **339**:339–343.

55. Jurczyk, M., Nowak, M., and Jankowska, E. (2002). Nanocrystalline LaNi$_{4-x}$Mn$_{0.75}$Al$_{0.25}$Co$_x$ electrode materials prepared by mechanical alloying (0≤x≤1.0). *J. Alloys Comp.* **340**:281–285.

56. Anani, A., Visintin, A., Petrov, K., and Srinivasan, S. (1994). Alloys for hydrogen storage in nickel/hydrogen and nickel/metal hydride batteries. *J. Power Sources* **47**:261–275.

57. Varin, R. A., Czujko, T., Wronski, Z. S. (2009) *Nanomaterials for Solid State Hydrogen Storage*, Springer.

58. Au, M., Pourarian, F., Simizu, S., Sankar, S. G., and Zhang, L. (1995). Electrochemical properties of TiMn$_2$-type alloys ball-milled with nickel powder. *J. Alloys Comp.* **223**:1–5.

59. Sun, D., Latroche, M., and Percheron-Guégan, A (1997). Activation behavior of mechanically Ni-coated Zr-based Laves phase hydride electrode. *J. Alloys Comp.* **257**:302–305.

60. Nohara, S., Inoue, H., Fukumoto, Y., and Iwakura, C. (1997). Effect of surface modification of an MgNi alloy with graphite by ball-milling on the rate of hydrogen absorption. *J. Alloys Comp.* **252**:L16–L18.

61. Jurczyk, M. (2004). Nanostructured electrode materials for Ni-MH$_x$ batteries prepared by mechanical alloying. *J. Mater. Sci.* **39**:5271–5274.

62. Bouaricha, S., Dodelet, J. P., Guay, D., Huot, J., and Schultz, R. (2001). Activation characteristics of graphite modified hydrogen absorbing materials. *J. Alloys Comp.* **325**:245–251.

63. Jurczyk, M., and Nowak, M. (2008). Nanomaterials for hydrogen storage synthesized by mechanical alloying, in: Eftekhari Ali (Ed.): *Nanostructured Materials in Electrochemistry* (Wiley), Chapter 9.

64. Linden, D. (2002). Thomas Reddy (ed.) *Handbook of Batteries Third Edition*, McGraw-Hill, Chapter 32, "Nickel Hydrogen Batteries".

65. Jankowska, E., and Jurczyk, M. (2004). Electrochemical properties of sealed Ni-MH batteries using nanocrystalline TiFe-type anodes. *J. Alloys Comp.* **372**:L9–L12.

66. Jurczyk, M., Jankowska, E., Nowak, M., Jakubowicz, J. (2002). Nanocrystalline titanium type metal hydrides prepared by mechanical alloying. *J. Alloys Comp.* **336**:265–269.

67. Skowronki, J. M., Sierczynska, A., and Kopczyk, M. (2002). Investigation of the influence of nickel content on the correlation between the hydrogen equilibrium pressure for hydrogen absorbing alloy and the capacity of MH electrodes in open and closed cells. *J. Solid State Electrochem.* **7**:11–16.

68. Majchrzycki, W. (2002). PhD Thesis: The influence of micro- and nano-structure on the electrochemical properties of electrode materials in Ni-MH$_x$ batteries, Poznan University of Technology.

69. Zhao, X. G., and Ma, L. Q. (2009). Recent progress in hydrogen storage alloys for nickel/metal hydride secondary batteries—review. *Int. J. Hydrogen Energy* **34**:4788–4796.

70. Young, K. H., and Nei, J. (2013). The current status of hydrogen storage alloy development for electrochemical applications, *Materials* **6**:4574–4608.

71. Yang, Q. M., Ciureanu, M., Ryan, D. H., and Strom-Olsen, J. O. (1998). Composite hydride electrode materials. *J. Alloys Comp.* **274**:266–273.

72. Pan, H. G., Liu, Y. F., Gao, M. X., Zhu, Y. F., and Lei, Y. Q. (2003). The structural and electrochemical properties of La$_{0.7}$Mg$_{0.3}$(Ni$_{0.85}$Co$_{0.15}$)$_x$ (x = 3.0–5.0) hydrogen storage alloys. *Int. J. Hydrogen Energy* **28**:1219–1228.

73. Zhao, X. Y., Ding, Yi., Yang, M., and Ma, L. Q. (2008). Effect of surface treatment on electrochemical properties of MmNi$_{3.8}$Co$_{0.75}$Mn$_{0.4}$Al$_{0.2}$ hydrogen storage alloy. *Int. J. Hydrogen Energy* **33**:81–86.

74. Tian, Q. F., Zhang, Y., Chu, H. L., Sun, L. X., Xu, F., Tan, Z. C., et al. (2006). The electrochemical performances of Mg$_{0.9}$Ti$_{0.1}$Ni$_{1-x}$Pd$_x$ (x = 0–0.15) hydrogen storage electrode alloys. *J. Power Sources* **159**:155–158.

75. Miao, H., Gao, M. X., Liu, Y. F., Lin, Y., Wang, J. H., and Pan, H. G. (2007). Microstructure and electrochemical properties of Ti-V-based multiphase hydrogen storage electrode alloys Ti$_{0.8}$Zr$_{0.2}$V$_{2.7}$Mn$_{0.5}$Cr$_{0.8-x}$Ni$_{1.265}$Fe$_x$ (x = 0.0–0.8). *Int. J. Hydrogen Energy* **32**: 3947–3953.

76. Young, K. (2016). Electrochemical applications of metal hydrides. Chapter 11 in *Compendium of Hydrogen Energy*. Elsevier Ltd. pp. 289–304.

Index

absorbed hydrogen 43, 45, 155, 169, 237, 284
AES, *see* Auger electron spectroscopy
AFM, *see* atomic force microscopy
alanates 48–50
alloy costs 43, 283
alloy electrodes 8, 158, 237, 266, 270
  cyclic stability of 266, 270
alloy formation 69
alloy phases
  metastable 294
  synthesize novel 67
alloy powders 46, 69, 162, 189, 212, 227, 289–290, 292
  amorphous parent 188, 190
  TiFe 135
alloying 44, 46, 69, 181, 202, 227, 231, 255, 261, 286, 289
  boride microplasma surface 30
alloys
  AB-type 45, 287
  AB_2-type 44, 287
  AB_5-type 43–44, 214, 283, 287
  amorphous 136, 165, 212, 243
  amorphous TiNi 165
  arc melted 159
  decrepitation of 43, 283
  Gd-substituted 270–271
  Ge-substituted 43, 283
  hydride-forming 42
  hydrogen-absorbing 138, 287
  LaNi 199

$LaNi_5$-type 46, 145, 288
magnesium-based 46, 227, 256, 289
magnesium-based hydrogen storage 43
mechanically alloyed 161, 163, 165, 167
metal hydride 128
metastable 156
Mg-based 237–238, 261
Mg_2Cu-type 234–235, 237
MgNi 143, 239
microcrystalline 203, 287–288
microcrystalline Mg_2Cu 235, 237
nanocrystalline AB_5 203
nanocrystalline $LaNi_{4.2}Al_{0.8}$ 74, 207–208
nanocrystalline Mg_2Ni-type 233–234, 251
nanocrystalline powder 7
nanocrystalline TiNi-type 145
over-stoichiometric 45, 284–285
polycrystalline $LaNi_{4.2}Al_{0.8}$ bulk 209
polycrystalline $TiNi_{0.75}Fe_{0.25}$ 144
TiFe-based hydrogen storage 131–132, 134, 136, 138, 140, 142, 144
TiFe-type 140, 284
$TiFe_{1-x}Ni_x$ 136
titanium 138, 154
unhydrided 46, 288

Yb-substituted 270–271
zirconium-based 192
ZrV$_2$-based hydrogen storage
   179–180, 182, 184, 186,
   188, 190, 192
aluminum atoms 73–74, 207
amorphization 71, 162, 170,
   262, 265
amorphous materials 71,
   133–134, 162–163, 170,
   181, 187, 204, 228, 246
arc melting 64, 159, 168, 189
argon atmosphere, high-purity
   204, 206, 228–229,
   234–235, 246
ASA, *see* atomic sphere
   approximation
atomic force microscopy (AFM)
   103–112, 114, 116, 164,
   184, 204, 229, 246, 267
  in hydrogen storage materials
     research 103–104, 106,
     108, 110, 112, 114, 116
atomic hydrogen 152, 158,
   282
atomic sphere approximation
   (ASA) 139, 214
Auger electron spectroscopy
   (AES) 73, 138, 207, 210
Auger electrons 73–74, 207,
   209, 211
Auger intensities 73–74, 207,
   209, 211

ball-milling 8, 67, 110, 132,
   165, 168, 228, 253–254
  high-energy 7, 63, 65, 68–69,
     184, 189, 286, 288, 294
ball mills, high-energy 63,
   66–68

batteries 44, 125–126, 179, 181,
   187, 279–280, 286, 288,
   291–292
  secondary 280
binder, metallic 157
borohydrides 48, 50
Bragg's Law 92–93
bulk alloy 152, 206–207
bulk microcrystalline LaNi
   209–210

carbon 25, 47, 73–74, 157, 207,
   209, 211
carbon dioxide emissions 1–2
carbon impurities, concentration
   of 74, 207, 210
carbon nanotubes 22, 47–48,
   110, 193
catalysts 2, 34, 49–50, 162, 165,
   184, 192–193, 272
CEC, *see* cyclic extrusion
   compression
charge-discharge cycles 231,
   236, 243, 247, 255
charging-discharging cycles 268
chemical vapor deposition
   (CVD) 17–18, 22, 24, 35
chromium 43–45, 88, 138–139,
   166–167, 182, 189, 202,
   256, 263, 271, 284–285
cobalt 159–160, 202, 286
composite materials 8, 22, 140,
   238
composites 109, 246–247, 249,
   253, 255
compounds, hydrogen-forming
   43, 45
copper 43–44, 88, 156–157,
   202, 218, 228, 234, 237,
   240–243, 245, 263, 271,
   283–285

crystallites 19, 95–97, 128, 134
CVD, *see* chemical vapor deposition
cyclic extrusion compression (CEC) 33

Debye–Scherrer camera 98–99
dehydrogenation kinetics 122, 154
diffraction 79, 92–93, 95, 97, 99
discharge capacities 135–136, 143–144, 152–154, 156–158, 160–161, 164–168, 187–189, 212–213, 231–232, 236, 242–243, 254–255, 270, 285–286, 288–289
  electrochemical 271
discharge kinetics 151–153

ECAP, *see* equal channel angular processing
electrochemical hydrogen absorption 264
electroconductivity 156
electrode alloy
  long life hydrogen storage 41
  low-cost RE–Mg–Ni-based 264
  optimized RE–Mg–Ni-based 272
  TiNi 271
electrodes
  metal hydride 184, 289
  metal-hydride 282
  Mg_2Cu 235–236
  nickel hydroxide 289
  TiNi 152, 167, 291
electromagnetic waves 85

electron beam lithography 26
electron beams 28, 136–137
electron transfer 156
electron–hole pairs 91
elemental metal powders 69
equal channel angular processing (ECAP) 33
equilibrium hydrogen pressure 186, 237

Fermi level 140–141, 144, 190, 217, 237, 239
friction stir processing (FSP) 33
FSP, *see* friction stir processing

gaseous hydrogen 44, 193, 283
gaseous hydrogenation properties 161, 169
grain boundaries 18–20, 84–85, 95, 133, 135, 163, 185, 188, 208, 211, 234, 237, 240, 286
graphene 25, 47–48
graphene-related materials 48
graphite 46–47, 140, 143, 218–219, 227, 238 239, 289

HEBM, *see* high-energy ball milling
high-energy ball milling (HEBM) 61, 63, 66–68, 184, 193, 246, 248, 266, 286
HVPE, *see* hydride vapor phase epitaxy
hydride formation energy 43, 283

hydride vapor phase epitaxy (HVPE) 23
hydrides 7–8, 42–44, 49, 133, 151, 154, 156, 166–167, 169, 202–203, 230, 265, 281–283, 288
  alloy 7
  binary 39–40
  intermetallic 40–41, 43, 45
  metal 5, 39–41, 46, 50, 61, 119–122, 124, 131, 266, 280, 289
  metallic 5, 40, 42
  unstable 49, 182, 265
hydrogen
  absorption-desorption process 42
  absorption/desorption properties 199–200
  absorption kinetics 233
  absorption of 131, 138, 162, 193, 200, 233, 283
  absorption properties 192
  adsorption of 48, 158
  atoms 3, 48, 163, 168, 188, 203, 289
  liquid 39–40
  redox reaction of 161
  sorption 7, 122, 155, 172
  sorption properties 7, 46, 48, 169, 193, 211, 267, 272
hydrogen capacity 49–50, 122, 151, 155, 200, 202–203, 211, 262, 286
hydrogen electrosorption 186
hydrogen energy 1, 5
hydrogen storage, gaseous phase 188
hydrogen storage alloys 39, 64, 158, 189–190, 192–193, 261–262, 264, 266, 268, 270, 272, 286, 288–289, 294–295

electrochemical properties of 189, 294–295
Mg-3d-based 227–228, 230, 232, 234, 236, 238, 240, 242, 244, 246, 248, 250, 252, 254, 256
hydrogen storage behavior 256
hydrogen storage materials, preparation of 63, 272
hydrogen storage nanomaterials 8, 41, 63, 67, 109, 116
hydrogen storage properties 119, 121–122, 252, 256, 262, 294
hydrogen storage systems 7, 61–62
hydrogenation 122, 124, 135, 138, 151, 154, 156, 158–159, 190, 192–193, 205, 208, 228–229, 233–235, 243–244
hydrogenation capacity 166, 169
hydrogenation kinetics 193, 203, 211
hydrolysis 33

interfaces, oxide–metal 73–74, 207, 209, 211
intermetallic compounds 7, 42, 228, 261
  stable 131, 288
intermetallic phases 199–200
ion gun etching system 234, 238, 240, 250
iron 43, 46, 132–133, 135, 139, 141–144, 166–167, 169, 200, 202, 240–243, 245, 255, 283–285, 288
iron impurities 74, 207

LaNi 8, 40–44, 62, 199–219,
    252–253, 262, 269–272,
    286–287, 295
LaNi$_5$-type compounds 201–202
LaNi$_5$-type phase 219
lanthanum atoms 73–74, 207,
    209, 211
Laue's equations 93, 95
Laves phases 165, 180–181, 295
liquid phase epitaxy (LPE) 23
lithography 17, 26, 35
LPE, see liquid phase epitaxy

MCAS, see mechano-chemical
    activation synthesis
mechano-chemical activation
    synthesis (MCAS) 68
melt spinning (MS) 64–65
MEMS, see micro electro
    mechanical systems
metal hydride electrode
    materials, electrochemical
    characterization of
    125–126, 128
metal hydride materials 42
metal hydrides
  hydriding-dehydriding
    properties of 61, 285
  hydriding-dehydriding
    properties of Mg-type 46,
    227, 289
  nanostructured 6–7, 211
metal-organic chemical vapor
    deposition (MOCVD) 17
metal organic vapor phase
    epitaxy (MOVPE) 24
metal plate 115
metastable phase 71, 132, 164
Mg$_{1.5}$Mn$_{0.5}$Ni alloy discharge
    capacities 232

Mg$_{1.5}$Mn$_{0.5}$Ni phase 252, 256
Mg$_2$Ni 41, 161, 228–230, 232,
    242, 246, 252, 271, 287
  amorphous 246, 248
Mg$_2$Ni alloys
  microcrystalline 46, 227, 230,
    289
  nanocrystalline 231, 252
micro electro mechanical
    systems (MEMS) 27, 152
microcrystalline alloy
    electrode 186
microcrystalline bulk sample
    208, 211
microcrystalline hydride
    materials 61, 64
microcrystalline LaNi 207, 210
microcrystalline Mg$_2$Cu 227,
    236–239
microcrystalline Mg$_2$Ni 233–234,
    249–250
MOCVD, see metal-organic
    chemical vapor deposition
MOVPE, see metal organic vapor
    phase epitaxy
MS, see melt spinning

nanocrystalline hydrides,
    development of 7–8
nanocrystalline ZrV$_2$-type
    materials, discharge
    capacities of 189
nanomaterials, synthesis
    methods 21, 23, 25, 27,
    29, 31, 33

physical vapor deposition 18, 28

sealed button cells, discharge capacities of 291–292

TiFe alloys
  discharge capacities of 135
  microcrystalline 138
TiNi alloys 110, 137, 149–152, 154–155, 159–161, 163–164, 166–169, 271
  alloyed 162, 165
  discharge capacity of 152
  electrochemical activation process in 154
  electrochemical properties of 160
  polycrystalline 144
TiNi materials, hydrogen storage properties of 149
TiO_2 27, 49, 133, 136, 156, 186
TiZrNi alloy 158

UHV, *see* ultra high vacuum
UHV annealing 249–250
ultra high vacuum (UHV) 24–25, 35, 210

valence band 91, 140, 144–145, 190, 207–208, 216, 237, 239, 246
  experimental XPS 144, 233, 237, 239–240, 249

water 2, 27, 33, 64
water electrolysis 2, 281
wear resistance 25, 28
welding, cold 67, 70–71

X-Ray Diffraction 79–80, 82, 84, 86, 88, 90, 92–94, 96, 98, 100, 181, 242
X-ray tubes 87–89, 97, 100
X-rays 79, 85–93, 97–98
XRD measurements 161–162, 170

zirconium 43–45, 165, 169, 179, 181–182, 189–193, 256, 263, 271, 283–285
ZrM_2-type alloy 45, 284
ZrV_2 8, 41, 43–44, 179–181, 184, 189–191, 193, 256, 287

PGSTL 11/01/2017